T0182031

Goldbach's Problem

Michael Th. Rassias

Goldbach's Problem

Selected Topics

Foreword by Jörg Brüdern and Preda Mihăilescu

Michael Th. Rassias
Institute of Mathematics
University of Zürich
Zürich
Switzerland

and

Program in Interdisciplinary Studies
Institute for Advanced Study
Princeton, NJ
USA

ISBN 978-3-319-57912-2 ISBN 978-3-319-57914-6 (eBook)
DOI 10.1007/978-3-319-57914-6

Library of Congress Control Number: 2017938559

Mathematics Subject Classification (2010): 11P32, 11P55, 40-XX, 41-XX, 26D15, 11-02, 26-02, 01AXX

Printed on acid-free paper

This Springer imprint is published by Springer Nature
The registered company is Springer International Publishing AG
The registered company address is: Gewerbestrasse 11, 6330 Cham, Switzerland

To my mother Ninetta and my sister Matina
for their love, patience and understanding.

Foreword

Number theory is reputed for presenting, more than any other mathematical discipline, problems that are often very easy to state, yet whose solution resisted over centuries. Diophantine equations are possibly the most popular example of difficult mathematical problems expressed in simple statements. Some, like Pell's equation $x^2 - ny^2 = 1$, whose solution by means of continued fractions was found by the very mathematician who proposed its study, became famous less for resisting over centuries as an open problem, but rather as a valuable object of study and teaching. Incidentally, the equation was proposed and solved by Lord Vincent Brouckner and not by John Pell, to whom Euler attributed the solution of this problem.
Most famous among diophantine equations, *Fermat's Last Theorem*, which states that $x^n + y^n = z^n$ admits no nontrivial integer solutions for $n > 2$, remained an open problem for more than 350 years, until Wiles's breakthrough in 1995. It is responsible for the development of important areas in algebraic number theory.
In analytic number theory, not all important questions have a simple statement—Riemann's Hypothesis being a good example of a fundamental conjecture, whose statement cannot be explained without assuming some background in complex analysis. Other questions, related to additive number theory, do have simple statements—such as for instance the question if there are infinitely many *twin primes* or infinitely many *Sophie Germain primes*, i.e. primes p such that either $p+2$ or $2p+1$ is also a prime. Another long-standing question of additive number theory is whether there are arithmetic progressions of arbitrary length consisting only of primes. The question was settled recently in the affirmative by B. Green and T. Tao using methods of combinatorial number theory.
The most famous problem of additive number theory remains however the famous conjecture of Goldbach, stating that every even number can be

written as the sum of two primes: since there is only one even prime, the same does certainly not hold for the odd numbers. Goldbach, a school teacher in Königsberg, had formulated the question in a letter to Euler dated 1742. The present book contains more details on the correspondence between Euler and Goldbach concerning the latter's conjecture. The question remains unsolved until today, despite some important results, like Vinogradov's theorem that every sufficiently large odd integer is the sum of three primes. The result was later improved by the Chinese School of number theory leading to the theorem of Chen Jingrun, which states that every large even number is the sum of either two primes or a prime and a product of two primes.

The attraction of Goldbach's conjecture to the interested layman is reflected in Apostolos Doxiadis's best-seller novel *Uncle Petros and Goldbach's Conjecture*, a novel which led the publisher of its American translation to offer a prize of one million dollars to whomever would prove the conjecture within two years after the year 2000. The prize has not been claimed!

Coming from the country of Doxiadis, the author of this book offers a detailed, self-contained yet easy to read introduction to some classical results on the Goldbach problem, to the attention of interested freshmen or any mathematically interested reader with basic background knowledge in analytic number theory. A winner at the age of 15 of the silver medal at the International Mathematical Olympiad in Japan, Michael Th. Rassias has cultivated his interest in number theory from an early age, developing the spirit and pleasure for explaining some precious jewels of this discipline to those of his age while using a minimal amount of prerequisites. He is the author of the delightful book *Problem-Solving and Selected Topics in Number Theory*, Springer, New York, 2011, a problem book which is filled with complete proofs of classical results in number theory.

The present book grew out of an essay that was part of the author's Part III studies at the University of Cambridge, for acquiring the Master of Advanced Study in Mathematics. Using the same detailed, step-by-step style like in the preceding *Problem-Solving* book, the author offers here a classical introduction to the Goldbach problem. This book introduces the uninitiated reader to the subject. A brief historical section is followed by a very detailed account of Vinogradov's important contribution: all large odd numbers are the sum of three primes. This result should be part of a course in analytic number theory. Here, the proof is presented in a step-by-step style that is accessible to anyone who has mastered elementary number theory. This part of the text also serves as a lure of the principles underlying the Hardy-Littlewood circle method, a technique that is the basic tool in areas such as diophantine analysis.

The third chapter of this book presents a new research result of the author with H. Maier on a version of Vinogradov's theorem: *under the assumption of the Generalized Riemann Hypothesis each sufficiently large odd integer can be expressed as the sum of a prime and two isolated primes*. Thus, this book could also be of interest to more advanced readers including researchers working on Goldbach's problem.

The final chapter discusses a combinatorial approach to Goldbach's conjecture due to Schnirelmann. This is not so common in the classroom today, but Schnirelmann's work has the interesting feature that it provides conclusions valid for all natural numbers, not only for large numbers.

This book contains also an Appendix presenting biographical remarks of some eminent mathematicians who have contributed in the subject, as well as a brief summary of Olivier Ramaré which sketches Helfgott's proof of Goldbach's ternary conjecture.

Given Rassias' love for detail and rigorous style, this book will be of certain interest to various audiences wishing to learn more about some of the fundamental techniques for challenging the Goldbach problem.

Jörg Brüdern
Preda Mihăilescu
Mathematics Institute
University of Göttingen
Göttingen, Germany

Acknowledgements

This monograph is essentially based on my Master's Essay which I had written in 2010–2011 under the supervision of Prof. Ben Green at the Department of Mathematics of the University of Cambridge. I would like to express my gratitude to Prof. Ben Green for proposing this very interesting topic and for his helpful guidance.

I am grateful to Profs. Jörg Brüdern and Preda Mihăilescu for their invaluable remarks and for writing the Foreword of this book. I am also grateful to Prof. Olivier Ramaré for contributing a brief article sketching Harald Helfgott's proof of Goldbach's Ternary Conjecture, which is featured at the Appendix of the present book. I would like to express my sincere thanks to Prof. Harald Helfgott who, in September 2013, invited me for a few days at École Normale Supérieure, Paris, and for investing some of his valuable time to discuss on the subject. I would also like to express my thanks to Prof. Robert C. Vaughan for contributing the paper *Goldbach's conjectures: A historical perspective* [65] to the volume *Open Problems in Mathematics*, which I edited jointly with Prof. John F. Nash. His paper was very useful in some parts of the present monograph.

I wish to offer my sincere thanks to Prof. Helmut Maier for reading the whole manuscript and for contributing essential remarks, which have helped to improve the presentation of the monograph. I would also like to thank Dr. Jakob Ditchen for providing useful comments.

Notwithstanding the fact that this monograph assumed its final form during the period I was conducting postdoctoral research at the Mathematics Department of Princeton University, it was basically composed during my first year of research towards a Ph.D. in Mathematics at ETH-Zürich. I thus

wish to express my gratitude to my Ph.D. advisor Prof. Emmanuel Kowalski for all his support and encouragement.
Finally, it is my pleasure to acknowledge the superb assistance of the staff of Springer for the publication of this monograph.

Michael Th. Rassias

Contents

Symbols

$e(x)$	Abbreviation for $e^{2\pi ix}$
$\Lambda(n)$	von Mangoldt function
$\mu(n)$	Möbius function
$\sigma_a(n)$	The sum of the ath powers of the positive divisors of n
$\tau(n)$	The number of positive divisors of n
$\phi(n)$	Euler phi function
$\pi(x)$	The number of primes not exceeding x
$f(x) = o(g(x))$	$\lim_{x\to+\infty} f(x)/g(x) = 0$, where $g > 0$
$f(x) = O(g(x))$	There exists a constant c, such that $\lvert f(x)\rvert < c\,g(x)$ for sufficiently large values of x
P_m	Denotes a product of at most m prime numbers
$c_n(m)$	Denotes the Ramanujan sum
$f(x) \ll g(x)$	If $f(x)$, $g(x) \geq 0$, then there exists a constant c, such that $\lvert f(x)\rvert < c\,g(x)$ for sufficiently large values of x
$f(x) \gg g(x)$	If $f(x)$, $g(x) \geq 0$, then there exists a constant c, such that $\lvert g(x)\rvert < c f(x)$ for sufficiently large values of x
(a, b)	The greatest common divisor of a and b
$\mathrm{lcm}\{a, b\}$	The least common multiple of a and b
$p^k \parallel n$	p^k divides n, but p^{k+1} does not divide n
$\lfloor x \rfloor$	The greatest integer not exceeding x
$\lceil x \rceil$	The least integer not less than x
	$[x] = \min_{k\in\mathbb{Z}} \lvert x - k\rvert$, where $x \in \mathbb{R}$

Introduction

In 1742 C. Goldbach, in two letters sent to L. Euler, formulated two conjectures. The first conjecture stated that every even integer can be represented as the sum of two prime numbers and the second one, that every integer greater than 2 can be represented as the sum of three prime numbers. Goldbach considered 1 to be a prime number, but since in modern mathematics 1 is not included in the set of primes, Goldbach's conjectures can be restated as follows:

Conjecture 1.0.1 (TERNARY GOLDBACH CONJECTURE—TGC)
Every odd integer greater than 5 can be represented as the sum of three prime numbers.

Conjecture 1.0.2 (BINARY GOLDBACH CONJECTURE—BGC)
Every even integer greater than 2 can be represented as the sum of two prime numbers.

It is amazing that, except for Euler's response to Goldbach, little is known about the interest of the mathematical community towards the proof of Goldbach's conjectures before 1900. It is also interesting to note that D. Hilbert in his lecture delivered before the International Congress of Mathematicians at Paris in 1900 (cf. [26]), in which he posed the 23 now famous problems, included Goldbach's conjecture as a part of Problem 8, entitled: "Problems of Prime Numbers".
Hilbert stated it as follows:
"After an exhaustive discussion of Riemann's prime number formula, perhaps we may sometime be in a position to attempt the rigorous solution of Goldbach's problem ... and further to attack the well-known question, whether

M.T. Rassias, *Goldbach's Problem*, DOI 10.1007/978-3-319-57914-6_1

there are an infinite number of pairs of prime numbers with the difference 2, or even the more general problem, whether the linear diophantine equation

$$ax + by + c = 0$$

(with given integral coefficients each prime to the others) is always solvable in prime numbers x and y."

1 Results Towards the Proof of the Ternary Goldbach Conjecture (TGC)

The TGC was successfully attacked for the first time in 1923 by G. H. Hardy and J. E. Littlewood. By applying their Circle Method and with the assumption that the Generalized Riemann Hypothesis holds true, they managed to prove that every sufficiently large odd integer can be represented as the sum of three prime numbers. More specifically, with the assumption of the Generalized Riemann Hypothesis, they proved that all but finitely many odd integers can be represented as the sum of three prime numbers (and that all but $O\left(x^{1/2+\epsilon}\right)$ even integers less than or equal to x can be represented as the sum of two prime numbers).

In 1926, B. Lucke proved in his doctoral dissertation that the term "sufficiently large" in the result of Hardy and Littlewood can be taken as 10^{32}. However, in 1937 I. M. Vinogradov by using the Circle Method eliminated the dependency on the Generalized Riemann Hypothesis and proved directly that all sufficiently large odd integers can be represented as the sum of three primes ([66, 67]).

Two years later, in 1939, K. G. Borodzkin, a student of Vinogradov proved that "sufficiently large" can be taken to mean greater than $3^{14348907}$. In 2002 though, M. C. Liu and T. Z. Wang proved that "sufficiently large" can be taken to mean greater than $2 \cdot 10^{1346}$.

A conditional proof of the TGC was presented in 1997 by J. M. Deshouillers, G. Effinger, H. te Riele and D. Zinoviev when they proved that the Generalized Riemann Hypothesis implies the TGC.

In 2012, T. Tao unconditionally proved that every odd integer larger than 1 is the sum of at most five prime numbers. The same year, T. Oliveira e Silva announced that he computationally verified Goldbach's Binary Conjecture for every even positive integer n with $2 < n \leq 4 \cdot 10^{18}$. However, this fact combined with some results by O. Ramaré–Y. Saouter and D. Platt yields that the TGC is true for every odd positive integer n with $5 < n \leq 1.23 \cdot 10^{27}$. In 2013, H. Helfgott and D. Platt improved this upper bound and showed that the

TGC is verified for every odd positive integer n with $5 < n \leq 8.875 \cdot 10^{30}$. But, a few months later, in May 2013, H. Helfgott presented an unconditional proof of the TGC [24].

2 Results Towards a Proof of the Binary Goldbach Conjecture (BGC)

Generally, the Hardy and Littlewood Circle Method is not particularly useful towards the proof of the BGC. Sieve methods turn out to be more powerful. The first mathematician to apply successfully a sieve method in order to attack the BGC was V. Brun. In 1919, he presented a sieve method, which was more effective than that of Eratosthenes (276 – 196 B.C.), based on the sieve function

$$S(A, P, r) = \sum_{\substack{a \in A \\ (a, P(r))=1}} 1 \,,$$

where A is a sequence of integers, P is an infinite set of prime numbers, r is a real number greater than 2 and

$$P(r) = \prod_{\substack{p \in P \\ p<r}} p \,.$$

In 1920, Brun managed with his method to prove that for every sufficiently large integer N, it follows that

$$2N = P_9 + P_9 \,,$$

where generally P_m denotes a product of at most m prime numbers.
Later, many mathematicians worked in the field of sieve methods. Among them are M. B. Barban, E. Bombieri, A. A. Buchstab, J. Chen, T. Estermann, P. X. Gallagher, H. Halberstam, H. Iwaniec, E. Kowalski, P. Kuhn, Y. V. Linnik, C. Pan, H. Rademacher, A. Rényi, H. -E. Richert, A. Selberg, A. I. Vinogradov and Y. Wang.
In 1951, Y. V. Linnik showed that there exists a constant C, such that every sufficiently large even integer can be represented as the sum of two prime numbers and at most C powers of 2.
In 2002, D. R. Heath-Brown and J. -C. Schlage-Puchta proved that $C = 13$ works for Linnik's result. In 2003, J. Pintz and I. Z. Ruzsa improved the

result of Heath-Brown and Schlage-Puchta by proving that $C = 8$ works for Linnik's result.
In 1947, A. Rényi proved that there exists a positive constant k such that, for every sufficiently large N, we have

$$2N = P_1 + P_k .$$

In 1957, Y. Wang proved that for every sufficiently large N, it follows that

$$2N = P_2 + P_3 .$$

Later, in 1966, J. Chen [7] announced (and published in 1973) the very important result that for every sufficiently large N, the following equality holds true:

$$2N = P_1 + P_2 .$$

Theorem 1.2.1 (CHEN'S THEOREM)
Let N be a sufficiently large even integer and

$$D(N) = \{p \ : \ p \leq N, \ N - p = P_2\} .$$

Then, the following inequality holds

$$|D(N)| > 0.67 \prod_{p>2} \left(1 - \frac{1}{(p-1)^2}\right) \prod_{\substack{p|N \\ p>2}} \frac{p-1}{p-2} \cdot \frac{N}{\log^2 N} .$$

In particular, every sufficiently large even integer N can be represented in the form $P_1 + P_2$.

Another remarkable result was presented in 1932 by L. G. Schnirelmann [58], when he proved that there exists a positive integer q, such that every integer greater than 1 can be represented as the sum of at most q prime numbers. In his work, he did not specify any possible value for the constant q. The first to determine an explicit value for q was K. I. Klimov, by proving in 1969 that one can take $q = 6 \cdot 10^9$. This result was substantially improved in 1972, when Klimov in collaboration with G. Z. Piltay and T. A. Sheptickaja, proved that q could be taken to be 115. Three years later, Klimov reduced this constant further to 55. In 1977, R. C. Vaughan proved in [62] that one could take $q = 27$. The same year, J. M. Deshouillers in [10] reduced this constant by one, namely proving that one can take $q = 26$. Finally, in 1983, H. Riesel and Vaughan proved in [56] that q can assume the value 19.
Furthermore, in 1938, N. G. Chudakov, J. G. van der Corput and T. Ester-

mann used a technique of Vinogradov and proved that the fraction of even integers which can be represented as the sum of two prime numbers tends asymptotically to 1.

Two years after the publication of Chen's result, in 1975, H. L. Montgomery and R. C. Vaughan [41] proved that there exist positive constants c_1, c_2, such that for all sufficiently large integers N, every even integer less than N can be represented as the sum of two prime numbers with at most $c_2 N^{1-c_1}$ exceptions.

In 1995, O. Ramaré proved in [47] that every even integer is in fact the sum of at most six prime numbers. For further results related to Goldbach's conjectures, the interested reader is referred to the Further Reading section as well as the Bibliography at the end of the book.

In this monograph, a step-by-step proof of Vinogradov's theorem is given, which states the following:

Theorem 1.2.2 (VINOGRADOV'S THEOREM)
There exists a natural number N, such that every odd positive integer n, with $n \geq N$, can be represented as the sum of three prime numbers.

We also present a recent result of H. Maier and M. Th. Rassias [37] on the representation of sufficiently large odd integers as a sum of a prime and two isolated primes. Additionally, we provide an outline of the basic steps of the proof of Schnirelmann's theorem. At the Appendix, an article of O. Ramaré which briefly sketches Helfgott's proof [24] of the TGC is presented as well.

Step-by-Step Proof of Vinogradov's Theorem

In the first section, we begin with some lemmas and theorems which will be useful in presenting a step-by-step proof of Vinogradov's theorem, which states that there exists a natural number N, such that every odd positive integer n, with $n \geq N$, can be represented as the sum of three prime numbers. The experienced reader may wish to skip this section.
In the second section, we present the Hardy-Littlewood Circle Method and describe the basic ideas which will be used in the presentation of the proof of Vinogradov's theorem.
The third, and most important section of this chapter, is devoted to Vaughan's proof of Vinogradov's theorem. A number of authors have presented Vaughan's proof in books and expositions. We also present here this proof, but in a step-by-step manner. More specifically, in the beginning of this section we describe in detail how the Circle Method can be applied to attack the Ternary Goldbach Conjecture (TGC). Firstly, we define the appropriate Major and Minor arcs, and afterwards, we investigate their contribution in the integral which describes the number of representations of an integer as the sum of three prime numbers.
The following references have been particularly useful in writing this chapter: [9, 14, 29, 33, 40, 43, 61, 64, 66, 67]. The paper of Vaughan [65] has been of exceptional interest in this monograph.

1 Introductory Lemmas and Theorems

In this section, we state some lemmas and theorems which will be useful during the step-by-step analysis of the proof of Vinogradov's theorem. These lemmas and theorems are essentially independent from each other. The following two lemmas can be easily verified, and thus, we omit the details.

M.T. Rassias, *Goldbach's Problem*, DOI 10.1007/978-3-319-57914-6_2

Lemma 1.1 *The following holds*

$$\sum_{l=1}^{d} e\left(\frac{ln}{d}\right) = \begin{cases} d, \text{ if } d \mid n \\ 0, \text{ if } d \nmid n, \end{cases}$$

where $e(x) = e^{2\pi ix}$, $x \in \mathbb{R}$.

Lemma 1.2 *Let* $a, b \in \mathbb{Z}$. *Then,*

$$\int_0^1 e(ax)e(-bx)dx = \begin{cases} 1, & \text{if } a = b \\ 0, & \text{if } a \neq b. \end{cases}$$

Definition 1.3 Let f be an arithmetic function. The series

$$D(f, s) = \sum_{n=1}^{+\infty} \frac{f(n)}{n^s},$$

where $s \in \mathbb{C}$, is called a Dirichlet series with coefficients $f(n)$.

We shall handle Dirichlet series for s being a real number.
Consider now a Dirichlet series, which is absolutely convergent for $s > s_0$.
If for these values of s it holds

$$\sum_{n=1}^{+\infty} \frac{f(n)}{n^s} = 0,$$

then $f(n) = 0$, for every integer n with $n \geq 1$.
If for these values of s it holds

$$D(f, s) = D(g, s),$$

then by the above argument it holds

$$f(n) = g(n), \text{ for every integer } n \text{ with } n \geq 1.$$

Theorem 1.4 *(1) Let* $D(f_1, s)$ *and* $D(f_2, s)$ *be convergent for* $s \in \mathbb{C}$. *Then the sum of* $D(f_1, s)$ *and* $D(f_2, s)$ *is obtained by*

$$D(f_1, s) + D(f_2, s) = \sum_{n=1}^{+\infty} \frac{f_1(n) + f_2(n)}{n^s}.$$

(2) Let $D(f_1, s)$ and $D(f_2, s)$ be absolutely convergent for $s \in \mathbb{C}$. Then the product of $D(f_1, s)$ and $D(f_2, s)$ is obtained by

$$D(f_1, s) \cdot D(f_2, s) = \sum_{n=1}^{+\infty} \frac{g(n)}{n^s},$$

where

$$g(n) = \sum_{n_1 n_2 = n} f_1(n_1) f_2(n_2).$$

Theorem 1.5 *Let f be a multiplicative function. Then, it holds*

$$D(f, s) = \prod_p \left(\sum_{n=0}^{+\infty} \frac{f(p^n)}{p^{ns}} \right),$$

where the product extends over all prime numbers p.

The basic idea of the proof of the theorem is the following:
It is true that

$$\prod_p \left(\sum_{n=0}^{+\infty} \frac{f(p^n)}{p^{ns}} \right) = \prod_p \left(\frac{f(1)}{1} + \frac{f(p)}{p^s} + \frac{f(p^2)}{p^{2s}} + \cdots \right)$$

$$= \left(\frac{f(1)}{1} + \frac{f(p_1)}{p_1^s} + \frac{f(p_1^2)}{p_1^{2s}} + \cdots \right) \left(\frac{f(1)}{1} + \frac{f(p_2)}{p_2^s} + \frac{f(p_2^2)}{p_2^{2s}} + \cdots \right) \cdots$$

$$= \sum \frac{f(p_1^{a_1}) \cdots f(p_k^{a_k})}{(p_1^{a_1} \cdots p_k^{a_k})^s}, \tag{1}$$

where[1] the sum extends over all possible combinations of multiples of powers of prime numbers. But, since the function $f(n)$ is multiplicative, it is evident that

$$\sum \frac{f(p_1^{a_1}) \cdots f(p_k^{a_k})}{(p_1^{a_1} \cdots p_k^{a_k})^s} = \sum \frac{f(p_1^{a_1} \cdots p_k^{a_k})}{(p_1^{a_1} \cdots p_k^{a_k})^s}$$

$$= \sum_{n=1}^{+\infty} \frac{f(n)}{n^s} = D(f, s).$$

□

[1] Here p_i denotes the ith prime number ($p_1 = 2, p_2 = 3, \ldots$).

Definition 1.6 The zeta function is defined by

$$\zeta(s) = \sum_{n=1}^{+\infty} \frac{1}{n^s},$$

for all real values of s with $s > 1$.

This function was defined for the first time in 1737 by Leonhard Euler (1707–1783). More than a century after Euler, in 1859, Georg Friedrich Bernhard Riemann (1826–1866) rediscovered the zeta function for complex values of s, while he was trying to prove the prime number Theorem.

Theorem 1.7 (EULER'S IDENTITY)

$$\zeta(s) = \prod_p \frac{1}{1 - p^{-s}}, \quad s \in \mathbb{R}, \ \textit{with } s > 1,$$

where the product extends over all prime numbers p.

The proof of Euler's Identity follows directly from Theorem 1.5.

Definition 1.8 Let $n \in \mathbb{N}$. The Möbius function $\mu(n)$ is defined as follows

$$\mu(n) = \begin{cases} 1, & \text{if } n = 1 \\ (-1)^k, & \text{if } n = p_1 p_2 \cdots p_k \text{ where } p_1, p_2, \ldots, p_k \text{ are } k \text{ distinct primes} \\ 0, & \text{in every other case} \end{cases}$$

Theorem 1.9

$$\sum_{d|n} \mu(d) = \begin{cases} 1, & \textit{if } n = 1 \\ 0, & \textit{if } n > 1, \end{cases}$$

where the sum extends over all positive divisors of the positive integer n.

Proof If $n = 1$ then the theorem obviously holds true, since by the definition of the Möbius function we know that $\mu(1) = 1$.
If $n > 1$ we can write

$$n = p_1^{a_1} p_2^{a_2} \cdots p_k^{a_k},$$

where $p_1, \ p_2, \ldots, \ p_k$ are distinct prime numbers.
Therefore

$$\sum_{d|n} \mu(d) = \mu(1) + \sum_{1 \le i \le k} \mu(p_i) + \sum_{\substack{i \ne j \\ 1 \le i,j \le k}} \mu(p_i p_j) + \cdots + \mu(p_1 p_2 \cdots p_k), \quad (1)$$

where generally the sum

$$\sum_{i_1 \neq i_2 \neq \cdots \neq i_\lambda} \mu(p_{i_1} p_{i_2} \cdots p_{i_\lambda})$$

extends over all possible products of λ distinct prime numbers. Hence, by (1) and the binomial identity, we obtain

$$\sum_{d|n} \mu(d) = 1 + \binom{k}{1}(-1) + \binom{k}{2}(-1)^2 + \cdots + \binom{k}{k}(-1)^k$$
$$= (1-1)^k = 0.$$

Therefore,

$$\sum_{d|n} \mu(d) = 0, \quad \textit{if} \quad n > 1.$$

□

Theorem 1.10 (The Möbius Inversion Formula) *Let $n \in \mathbb{N}$. If*

$$g(n) = \sum_{d|n} f(d)$$

then

$$f(n) = \sum_{d|n} \mu\left(\frac{n}{d}\right) g(d).$$

The converse also holds.

Proof For every arithmetic function $m(n)$, it holds

$$\sum_{d|n} m(d) = \sum_{d|n} m\left(\frac{n}{d}\right).$$

Therefore, it is evident that

$$\sum_{d|n} \mu\left(\frac{n}{d}\right) g(d) = \sum_{d|n} \mu(d) g\left(\frac{n}{d}\right). \tag{1}$$

But

$$\sum_{d|n} \mu(d) g\left(\frac{n}{d}\right) = \sum_{d|n} \left(\mu(d) \cdot \sum_{\lambda | \frac{n}{d}} f(\lambda) \right).$$

Hence, we get

$$\sum_{d|n} \mu(d) g\left(\frac{n}{d}\right) = \sum_{\lambda d|n} \mu(d) f(\lambda) .$$

Similarly,

$$\sum_{\lambda|n} \left(f(\lambda) \cdot \sum_{d|\frac{n}{\lambda}} \mu(d) \right) = \sum_{\lambda d|n} \mu(d) f(\lambda) .$$

Thus,

$$\sum_{d|n} \mu(d) g\left(\frac{n}{d}\right) = \sum_{\lambda|n} \left(f(\lambda) \cdot \sum_{d|\frac{n}{\lambda}} \mu(d) \right) . \tag{2}$$

However, by Theorem 1.9 we get

$$\sum_{d|\frac{n}{\lambda}} \mu(d) = 1 \text{ if and only if } \frac{n}{\lambda} = 1 ,$$

and in every other case, the sum is equal to zero. Thus, for $n = \lambda$ we obtain

$$\sum_{\lambda|n} \left(f(\lambda) \cdot \sum_{d|\frac{n}{\lambda}} \mu(d) \right) = f(n). \tag{3}$$

Therefore, by (1), (2) and (3) it follows that if

$$g(n) = \sum_{d|n} f(d)$$

then

$$f(n) = \sum_{d|n} \mu\left(\frac{n}{d}\right) g(d) .$$

Conversely we shall prove that if

$$f(n) = \sum_{d|n} \mu\left(\frac{n}{d}\right) g(d) ,$$

then

$$g(n) = \sum_{d|n} f(d).$$

We have

$$\begin{aligned}\sum_{d|n} f(d) &= \sum_{d|n} f\left(\frac{n}{d}\right) \\ &= \sum_{d|n} \sum_{\lambda|\frac{n}{d}} \mu\left(\frac{n}{\lambda d}\right) g(\lambda) \\ &= \sum_{d\lambda|n} \mu\left(\frac{n}{\lambda d}\right) g(\lambda) \\ &= \sum_{\lambda|n} g(\lambda) \sum_{d|\frac{n}{\lambda}} \mu\left(\frac{n}{\lambda d}\right)\end{aligned}$$

The sum

$$\sum_{d|\frac{n}{\lambda}} \mu\left(\frac{n}{\lambda d}\right) = 1$$

if and only if $n = \lambda$, and in every other case, it is equal to zero. Hence, for $n = \lambda$ we obtain

$$\sum_{d|n} f(d) = g(n).$$

This completes the proof of the theorem. □

Theorem 1.11 *For $s > 1$,*

$$\frac{1}{\zeta(s)} = \sum_{n=1}^{+\infty} \frac{\mu(n)}{n^s}.$$

Proof By Euler's identity, we have

$$\frac{1}{\zeta(s)} = \prod_p \left(1 - \frac{1}{p^s}\right) \tag{2.1}$$

Additionally, by Theorem 1.5, it is evident that

$$\sum_{n=1}^{+\infty} \frac{\mu(n)}{n^s} = \prod_p \left(1 - \frac{1}{p^s}\right). \tag{2.2}$$

By the above formulas (2.1) and (2.2), the theorem follows. □

Definition 1.12 Let $n \in \mathbb{N}$. The Von Mangoldt function $\Lambda(n)$ is defined as follows

$$\Lambda(n) = \begin{cases} \log p, & \text{if } n = p^k \text{ for some prime number } p \text{ and some } k \in \mathbb{N} \\ 0, & \text{otherwise.} \end{cases}$$

Theorem 1.13 *Let $n \in \mathbb{N}$. Then, it holds*

$$\sum_{d|n} \Lambda(d) = \log n \tag{L1}$$

and

$$\Lambda(n) = -\sum_{d|n} \mu(d) \log d. \tag{L2}$$

Proof It is evident that the theorem holds true in the case when $n = 1$. Hence, let us assume that $n > 1$. If

$$n = p_1^{a_1} p_2^{a_2} \dots p_m^{a_m}$$

is the cannonical representation of n by its prime factors, we get

$$\begin{aligned} \log n &= \sum_{i=1}^{m} a_i \log p_i \\ &= \sum_{i=1}^{m} \sum_{q=1}^{a_i} \log p_i \\ &= \sum_{i=1}^{m} \sum_{q=1}^{a_i} \Lambda(p_i^q) \\ &= \sum_{d|n} \Lambda(d). \end{aligned}$$

This completes the proof of (L1).
By the Möbius inversion formula and (L1) we obtain

$$\begin{aligned} \Lambda(n) &= \sum_{d|n} \mu(d) \log \frac{n}{d} \\ &= \log n \sum_{d|n} \mu(d) - \sum_{d|n} \mu(d) \log d. \end{aligned}$$

Therefore, by Theorem 1.9 and the above formula, (L2) follows. □

Theorem 1.14 *For $s > 1$, it holds*

$$\sum_{n=1}^{+\infty} \frac{\Lambda(n)}{n^s} = -\frac{1}{\zeta(s)} \frac{d\zeta(s)}{ds}.$$

Proof From the definition of the Riemann zeta function and a well-known theorem of termwise differentiation of an infinite series, it follows that

$$\frac{d\zeta(s)}{ds} = -\sum_{n=1}^{+\infty} \frac{\log n}{n^s}. \tag{1}$$

Additionally, by Theorem 1.11, we know that

$$\frac{1}{\zeta(s)} = \sum_{n=1}^{+\infty} \frac{\mu(n)}{n^s}. \tag{2}$$

By Theorem 1.4, if we multiply (1), (2), we obtain

$$\frac{1}{\zeta(s)} \frac{d\zeta(s)}{ds} = -\sum_{n=1}^{+\infty} \frac{g(n)}{n^s} \tag{3}$$

where

$$g(n) = \sum_{n_1 n_2 = n} \mu(n_1) \log n_2.$$

But, by Theorem 1.13, it follows that

$$g(n) = \Lambda(n).$$

Therefore, by (3) and the above relation, the theorem follows. □

Definition 1.15 The exponential sum

$$c_n(m) = \sum_{\substack{1 \le q \le n \\ (q,n)=1}} e\left(\frac{qm}{n}\right),$$

is called the Ramanujan sum $c_n(m)$.

Note By the use of the Chinese Remainder Theorem (cf. [55]) it can be proved that the Ramanujan sum $c_n(m)$ is a multiplicative function of n.

Lemma 1.16 *For the Ramanujan sum $c_n(m)$, we have*

$$c_n(m) = \sum_{d|(n,m)} \mu\left(\frac{n}{d}\right) d\,.$$

Proof By Theorem 1.9, we know that for the Möbius function it holds

$$\sum_{d|n} \mu(d) = \begin{cases} 1, & \text{if } n = 1 \\ 0, & \text{if } n > 1 \end{cases}$$

Therefore, we can write

$$\begin{aligned} c_n(m) &= \sum_{\substack{1 \le q \le n \\ (q,n)=1}} e\left(\frac{qm}{n}\right) \\ &= \sum_{q=1}^{n} \left(e\left(\frac{qm}{n}\right) \sum_{d|(q,n)} \mu(d) \right) \\ &= \sum_{d|n} \sum_{k=1}^{n/d} e\left(\frac{km}{n/d}\right) \mu(d) \\ &= \sum_{d|n} \sum_{k=1}^{d} e\left(\frac{km}{d}\right) \mu\left(\frac{n}{d}\right) . \end{aligned}$$

However, by Lemma 1.1, we obtain

$$c_n(m) = \sum_{\substack{d|n \\ d|m}} \mu\left(\frac{n}{d}\right) d\,.$$

But, $d \mid n$ and $d \mid m$ is equivalent to $d \mid (n, m)$. Hence, it follows that

$$c_n(m) = \sum_{d|(n,m)} \mu\left(\frac{n}{d}\right) d\,.$$

□

Lemma 1.17 *Let x be a real number. Then,*

$$\left| \sum_{n=B_1+1}^{B_2} e(xn) \right| \le \min\left\{ \frac{1}{[x]}, B_2 - B_1 \right\} ,$$

where B_1, B_2 are integers with $B_1 < B_2$ and $\left[y\right] = \min_{k\in\mathbb{Z}} |y-k|$, where $y \in \mathbb{R}$.

Proof Let us suppose that x is not an integer. In this case, we have

$$\left|\sum_{n=B_1+1}^{B_2} e(xn)\right| \leq \frac{\left|e^{2\pi i(B_1+1)x}\right| + \left|e^{2\pi i(B_2+1)x}\right|}{\left|e^{2\pi i(x/2)} - e^{2\pi i(-x/2)}\right|}$$

$$= \frac{2}{\left|e^{\pi ix} - e^{-\pi ix}\right|}\,.$$

But

$$i\sin\theta = \frac{e^{i\theta} - e^{-i\theta}}{2}$$

and therefore, we obtain

$$\left|\sum_{n=B_1+1}^{B_2} e(xn)\right| \leq \frac{1}{|2\sin(\pi x)|}\,.$$

In addition, it is true that

$$\pi x = \pi(k \pm [x])\,,$$

where k is the nearest integer to x. Thus, it follows that

$$\frac{1}{|2\sin(\pi x)|} = \frac{1}{2\sin(\pi\,[x])}\,.$$

However, it is a well-known fact that

$$\frac{\sin\theta}{\theta} > \frac{2}{\pi}\,,\ \text{for} \ -\frac{\pi}{2} < \theta < \frac{\pi}{2},\ \theta \neq 0\,.$$

Hence, for $\theta = \pi\,[x] > 0$, we get

$$\sin\pi\,[x] > \frac{2\pi\,[x]}{\pi} = 2\,[x]\,.$$

From all the above, it is evident that

$$\left|\sum_{n=B_1+1}^{B_2} e(xn)\right| \leq \frac{1}{[x]}\,. \qquad (1)$$

Of course, it is clear that

$$\left|\sum_{n=B_1+1}^{B_2} e(xn)\right| \leq \sum_{n=B_1+1}^{B_2} |e(xn)| = B_2 - B_1 \,. \tag{2}$$

Hence, from (1) and (2), we obtain

$$\left|\sum_{n=B_1+1}^{B_2} e(xn)\right| \leq \min\left\{\frac{1}{[x]}, B_2 - B_1\right\} .$$

□

Theorem 1.18 *Let $\tau(n)$ denote the divisor function, defined by*

$$\tau(n) = \sum_{\substack{d|n \\ d\geq 1}} 1 \,.$$

Then, for any real number x, with $x \geq 2$, it holds

$$\sum_{n\leq x} \tau^2(n) \ll x \log^3 x \,.$$

Proof We have

$$\begin{aligned}\sum_{n\leq x} \tau^2(n) &= \sum_{n\leq x}\left(\sum_{d_1|n} 1\right)\left(\sum_{d_2|n} 1\right)\\ &= \sum_{d_1,d_2\leq x} \sum_{\substack{n\leq x \\ n\equiv 0 \pmod{\mathrm{lcm}\{d_1,d_2\}}}} 1 \,.\end{aligned}$$

We write $e = (d_1, d_2)$, $d_1 = t_1 e$, $d_2 = t_2 e$.
We have $\mathrm{lcm}\{d_1, d_2\} = t_1 t_2 e$ and thus

$$\begin{aligned}\sum_{n\leq x} \tau^2(n) &\leq \sum_{e\leq x} \sum_{t_1,\, t_2\leq x} \left\lfloor \frac{x}{t_1 t_2 e} \right\rfloor\\ &\leq x\left(\sum_{e\leq x} \frac{1}{e}\right)^3\\ &\ll x \log^3 x.\end{aligned}$$

This completes the proof of the theorem. □

Theorem 1.19 (LEGENDRE'S THEOREM)
The largest power of p which divides the integer n! is[2]

$$\sum_{k=1}^{+\infty} \left\lfloor \frac{n}{p^k} \right\rfloor$$

Proof The number of factors of $n!$ which are divisible by p, is $\lfloor n/p \rfloor$. More specifically, these factors are the integers:

$$1 \cdot p,\ 2 \cdot p, \ldots, \left\lfloor \frac{n}{p} \right\rfloor \cdot p\,.$$

However, some factors of $n!$ are divisible by at least the second power of p, namely they contain p^2 at least one time. These factors are the integers:

$$1 \cdot p^2,\ 2 \cdot p^2, \ldots, \left\lfloor \frac{n}{p^2} \right\rfloor \cdot p^2\,,$$

which are exactly

$$\left\lfloor \frac{n}{p^2} \right\rfloor$$

in number.
If we continue similarly for higher powers of p, it follows that the integer $n!$ contains the prime number p exactly

$$\left\lfloor \frac{n}{p} \right\rfloor + \left\lfloor \frac{n}{p^2} \right\rfloor + \cdots + \left\lfloor \frac{n}{p^k} \right\rfloor + \cdots$$

times and therefore that is exactly the largest power of p which divides $n\,!$.
The above sum is finite since for $k > r$, where $p^r \geq n$, it holds

$$\left\lfloor \frac{n}{p^k} \right\rfloor = 0.$$

□

Definition 1.20 We define $\pi(x)$ to be the number of primes which do not exceed a given real number x.

---

[2]By $\lfloor x \rfloor$ we denote the integer part of x and by $\lceil x \rceil$ the least integer, greater than or equal to x.

Theorem 1.21 (Chebyshev's Inequality)
For every positive integer n, where $n \geq 2$, the following inequality holds

$$\frac{1}{6} \cdot \frac{n}{\log n} < \pi(n) < 6 \cdot \frac{n}{\log n}$$

Proof We claim that

$$2^n \leq \binom{2n}{n} < 4^n . \tag{1}$$

The inequality

$$2^n \leq \binom{2n}{n}$$

follows by mathematical induction.
For $n = 2$ one has

$$4 \leq \binom{4}{2} = 6 ,$$

which holds true. Suppose that (1) is valid for n, i.e.

$$2^n \leq \binom{2n}{n} .$$

It suffices to prove (1) for $n + 1$.
It is clear that

$$\begin{aligned}\binom{2n+2}{n+1} &= \frac{(2n+2)!}{(n+1)!(n+1)!} \\ &= \frac{(2n)!}{n\,!\,n\,!}\frac{(2n+1)(2n+2)}{(n+1)^2} \\ &\geq 2^n \frac{(2n+1)(2n+2)}{(n+1)^2} .\end{aligned}$$

It is enough to prove that

$$\frac{(2n+1)(2n+2)}{(n+1)^2} \geq 2\,, \text{for } n \geq 2\,.$$

However,

$$\frac{(2n+1)(2n+2)}{(n+1)^2} \geq 2 \Leftrightarrow 2n \geq 0\,,$$

which clearly holds true.
Thus,

$$\binom{2n+2}{n+1} \geq 2^{n+1}$$

and therefore, we have proved that

$$2^n \leq \binom{2n}{n}$$

for every positive integer n, where $n \geq 2$.
The proof of the right-hand side of inequality (1) follows from the fact that

$$\binom{2n}{n} < \binom{2n}{0} + \binom{2n}{1} + \cdots + \binom{2n}{2n} = 2^{2n} = 4^n\,.$$

From (1), we get that

$$\log 2^n \leq \log \frac{(2n)!}{n!n!} < \log 4^n$$

and therefore

$$n \log 2 \leq \log(2n)! - 2\log n! < n \log 4\,. \tag{2}$$

However, from Legendre's Theorem (see Theorem 1.19), it follows that

$$n! = \prod_{p \leq n} p^{j(n,p)}\,, \tag{3}$$

where

$$j(n,p) = \sum_{k=1}^{+\infty} \left\lfloor \frac{n}{p^k} \right\rfloor\,.$$

From (3), we obtain

$$\log n! = \log \prod_{p \le n} p^{j(n,p)}$$
$$= \sum_{p \le n} \log p^{j(n,p)}$$
$$= \sum_{p \le n} j(n,p) \log p .$$

By applying this result, we get

$$\log(2n)! - 2\log n! = \sum_{p \le 2n} j(n,p)\log p - 2\sum_{p \le n} j(n,p)\log p$$
$$= \sum_{p \le 2n} \left(\sum_{k=1}^{+\infty} \left\lfloor \frac{2n}{p^k} \right\rfloor \right) \log p - 2 \sum_{p \le n} \left(\sum_{k=1}^{+\infty} \left\lfloor \frac{n}{p^k} \right\rfloor \right) \log p .$$

However,

$$\sum_{p \le n} \left(\sum_{k=1}^{+\infty} \left\lfloor \frac{n}{p^k} \right\rfloor \right) \log p = \sum_{p \le 2n} \left(\sum_{k=1}^{+\infty} \left\lfloor \frac{n}{p^k} \right\rfloor \right) \log p$$

since for $p > n$ it is true that

$$\left\lfloor \frac{n}{p^k} \right\rfloor = 0 .$$

Therefore,

$$\log(2n)! - 2\log n! = \sum_{p \le 2n} \left(\sum_{k=1}^{+\infty} \left\lfloor \frac{2n}{p^k} \right\rfloor - 2 \sum_{k=1}^{+\infty} \left\lfloor \frac{n}{p^k} \right\rfloor \right) \log p$$
$$= \sum_{p \le 2n} \left[\sum_{k=1}^{+\infty} \left(\left\lfloor \frac{2n}{p^k} \right\rfloor - 2 \left\lfloor \frac{n}{p^k} \right\rfloor \right) \right] \log p .$$

However, it holds

$$\left\lfloor \frac{2n}{p^k} \right\rfloor - 2 \left\lfloor \frac{n}{p^k} \right\rfloor < \frac{2n}{p^k} - 2\left(\frac{n}{p^k} - 1 \right) = 2 .$$

Thus, clearly

$$\left\lfloor \frac{2n}{p^k} \right\rfloor - 2\left\lfloor \frac{n}{p^k} \right\rfloor = 0 \text{ or } 1 .$$

The terms of the infinite summation

$$\sum_{k=1}^{+\infty} \left(\left\lfloor \frac{2n}{p^k} \right\rfloor - 2\left\lfloor \frac{n}{p^k} \right\rfloor \right)$$

assume the value zero for k such that $p^k > 2n$, that means for

$$k > \frac{\log 2n}{\log p} .$$

Thus,

$$\begin{aligned}\sum_{k=1}^{+\infty} \left(\left\lfloor \frac{2n}{p^k} \right\rfloor - 2\left\lfloor \frac{n}{p^k} \right\rfloor \right) &= \sum_{k=1}^{\left\lfloor \frac{\log 2n}{\log p} \right\rfloor} \left(\left\lfloor \frac{2n}{p^k} \right\rfloor - 2\left\lfloor \frac{n}{p^k} \right\rfloor \right) \\ &\le \sum_{k=1}^{\left\lfloor \frac{\log 2n}{\log p} \right\rfloor} 1 .\end{aligned}$$

Hence,

$$\begin{aligned}\log(2n)! - 2\log n! &\le \sum_{p \le 2n} \left(\sum_{k=1}^{\left\lfloor \frac{\log 2n}{\log p} \right\rfloor} 1 \right) \log p \\ &\le \sum_{p \le 2n} \frac{\log 2n}{\log p} \log p \\ &= \sum_{p \le 2n} \log 2n \\ &= \pi(2n) \log 2n .\end{aligned}$$

From this relation and inequality (2), it follows that

$$n \log 2 \le \pi(2n) \log 2n$$

$$\Leftrightarrow \pi(2n) \geq \frac{n \log 2}{\log 2n} > \frac{n/2}{\log 2n} = \frac{2n}{4 \log 2n}$$

$$\Leftrightarrow \pi(2n) > \frac{1}{4} \cdot \frac{2n}{\log 2n} > \frac{1}{6} \cdot \frac{2n}{\log 2n}. \tag{4}$$

Therefore, the inequality

$$\frac{1}{6} \cdot \frac{n}{\log n} < \pi(n)$$

is satisfied if n is an even integer. It remains to examine the case where n is an odd integer.
It is true that

$$\begin{aligned}\pi(2n+1) \geq \pi(2n) &> \frac{1}{4} \cdot \frac{2n}{\log 2n} \\ &= \frac{1}{4} \cdot \frac{2n}{2n+1} \cdot \frac{2n+1}{\log 2n} \\ &> \frac{1}{4} \cdot \frac{2n}{2n+1} \cdot \frac{2n+1}{\log(2n+1)}.\end{aligned}$$

It is evident that

$$\frac{2n}{2n+1} \geq \frac{2}{3}$$

for every positive integer n.
Therefore

$$\begin{aligned}\pi(2n+1) &> \frac{1}{4} \cdot \frac{2}{3} \cdot \frac{2n+1}{\log(2n+1)} \\ &= \frac{1}{6} \cdot \frac{2n+1}{\log(2n+1)}\end{aligned}$$

Hence, the inequality

$$\frac{1}{6} \cdot \frac{n}{\log n} < \pi(n)$$

is also satisfied in the case where n is an odd integer.
Thus

$$\frac{1}{6} \cdot \frac{n}{\log n} < \pi(n),$$

for every positive integer n, with $n \geq 2$.

We will now prove the inequality

$$\pi(n) < 6 \cdot \frac{n}{\log n}$$

for every positive integer n with $n \geq 2$.
We have already proved that

$$\log(2n)! - 2\log n! = \sum_{p \leq 2n} \left[\sum_{k=1}^{+\infty} \left(\left\lfloor \frac{2n}{p^k} \right\rfloor - 2 \left\lfloor \frac{n}{p^k} \right\rfloor \right) \right] \log p \ ,$$

where none of the terms

$$\left\lfloor \frac{2n}{p^k} \right\rfloor - 2 \left\lfloor \frac{n}{p^k} \right\rfloor$$

is negative.
Therefore, it is clear that

$$\left\lfloor \frac{2n}{p} \right\rfloor - 2 \left\lfloor \frac{n}{p} \right\rfloor \leq \sum_{k=1}^{+\infty} \left(\left\lfloor \frac{2n}{p^k} \right\rfloor - 2 \left\lfloor \frac{n}{p^k} \right\rfloor \right) .$$

Thus

$$\begin{aligned} \log(2n)! - 2\log n! &\geq \sum_{p \leq 2n} \left(\left\lfloor \frac{2n}{p} \right\rfloor - 2 \left\lfloor \frac{n}{p} \right\rfloor \right) \log p \\ &\geq \sum_{n < p \leq 2n} \left(\left\lfloor \frac{2n}{p} \right\rfloor - 2 \left\lfloor \frac{n}{p} \right\rfloor \right) \log p \ . \end{aligned}$$

However for the prime numbers p, such that $n < p \leq 2n$ one has

$$\left\lfloor \frac{2n}{p} \right\rfloor - 2 \left\lfloor \frac{n}{p} \right\rfloor = 1 \ ,$$

since

$$\left\lfloor \frac{2n}{p} \right\rfloor = 1 \text{ and } \left\lfloor \frac{n}{p} \right\rfloor = 0 \ .$$

Hence,

$$\log(2n)! - 2\log n! \geq \sum_{n < p \leq 2n} \log p \ . \tag{5}$$

By the definition of Chebyshev's function $\vartheta(x)$, one has

$$\vartheta(x) = \sum_{p \le x} \log p \ .$$

Therefore, (5) can be written as follows:

$$\log(2n)! - 2\log n! \ge \vartheta(2n) - \vartheta(n) \, .$$

Thus by meansof (2), we obtain

$$\vartheta(2n) - \vartheta(n) < n \log 4 \, . \tag{6}$$

Suppose that the positive integer n can be expressed as an exact power of 2. Then from (6), it follows

$$\vartheta(2 \cdot 2^m) - \vartheta(2^m) < 2^m \log 2^2$$

and therefore

$$\vartheta(2^{m+1}) - \vartheta(2^m) < 2^{m+1} \log 2 \ .$$

For $m = 1, \ 2, \ \dots \ , \ \lambda - 1, \ \lambda$ the above inequality, respectively, yields

$$\left.\begin{array}{rr} \vartheta(2^2) - \vartheta(2) < & 2^2 \log 2 \\ \vartheta(2^3) - \vartheta(2^2) < & 2^3 \log 2 \\ \vdots & \\ \vartheta(2^\lambda) - \vartheta(2^{\lambda-1}) < & 2^\lambda \log 2 \\ \vartheta(2^{\lambda+1}) - \vartheta(2^\lambda) < & 2^{\lambda+1} \log 2 \end{array}\right\}$$

Adding up the above inequalities, we get

$$\vartheta(2^{\lambda+1}) - \vartheta(2) < (2^2 + 2^3 + \ldots + 2^\lambda + 2^{\lambda+1}) \log 2 \ .$$

But $\vartheta(2) = \log 2$, therefore

$$\begin{aligned} \vartheta(2^{\lambda+1}) &< (1 + 2^2 + 2^3 + \ldots + 2^\lambda + 2^{\lambda+1}) \log 2 \\ &= (2^{\lambda+2} - 1) \log 2 \ . \end{aligned}$$

Hence

$$\vartheta(2^{\lambda+1}) < 2^{\lambda+2} \log 2 \ . \tag{7}$$

For every positive integer n we can choose a suitable integer m such that

$$2^m \leq n \leq 2^{m+1} .$$

Then

$$\vartheta(n) = \sum_{p \leq n} \log p \leq \sum_{p \leq 2^{m+1}} \log p = \vartheta(2^{m+1})$$

and by means of (7), it follows that

$$\vartheta(n) < 2^{m+2} \log 2 = 2^2 \cdot 2^m \log 2 \leq 4n \log 2 . \tag{8}$$

Let N be the number of primes p_i, such that

$$n^r < p_i \leq n$$

where $0 < r < 1$, for $i = 1, 2, \cdots, N$. Then

$$\left.\begin{array}{ll} \log n^r < & \log p_1 \\ \log n^r < & \log p_2 \\ \vdots & \\ \log n^r < & \log p_N \end{array}\right\} \Rightarrow N \log n^r < \sum_{n^r < p \leq n} \log p .$$

and therefore

$$(\pi(n) - \pi(n^r)) \log n^r < \sum_{n^r < p \leq n} \log p . \tag{9}$$

It is obvious that

$$\vartheta(n) \geq \sum_{n^r < p \leq n} \log p . \tag{10}$$

Therefore by means of (8), (9) and (10), one has

$$\begin{aligned} (\pi(n) - \pi(n^r)) \log n^r &< 4n \log 2 \\ \Leftrightarrow \pi(n) \log n^r &< 4n \log 2 + \pi(n^r) \log n^r \\ \Leftrightarrow \pi(n) &< \frac{4n \log 2}{\log n^r} + \pi(n^r) \\ &< \frac{4n \log 2}{r \log n} + n^r . \end{aligned}$$

Thus, equivalently, we obtain

$$\pi(n) < \frac{n}{\log n}\left(\frac{4\log 2}{r} + n^{r-1}\log n\right). \tag{11}$$

Consider the function defined by the formula

$$f(x) = \frac{\log x}{x^{1-r}}, \quad x \in \mathbb{R}^+ .$$

Then

$$f'(x) = \frac{\frac{1}{x}x^{1-r} - (1-r)x^{-r}\log x}{(x^{1-r})^2} .$$

It is clear that

$$f'(x) = 0$$

if

$$x^{-r} = (1-r)x^{-r}\log x \Leftrightarrow \log x = \frac{1}{1-r},$$

that means

$$x = e^{1/(1-r)} .$$

For $x = e^{1/(1-r)}$ the function $f(x)$ attains its maximal value.
Thus

$$f(x) \le \frac{1}{e(1-r)} \Rightarrow f(n) \le \frac{1}{e(1-r)},$$

and therefore

$$n^{r-1}\log n \le \frac{1}{e(1-r)} . \tag{12}$$

From (11) and (12), it follows

$$\pi(n) < \frac{n}{\log n}\left(\frac{4\log 2}{r} + \frac{1}{e(1-r)}\right).$$

Set $r = \frac{2}{3}$. Then

$$\pi(n) < \frac{n}{\log n}\left(6\log 2 + \frac{3}{e}\right) .$$

However, it holds

$$6\log 2 + \frac{3}{e} < 6 \text{ and thus } \pi(n) < 6 \cdot \frac{n}{\log n}$$

Hence for every positive integer n, where $n \geq 2$, the following inequality holds

$$\frac{1}{6} \cdot \frac{n}{\log n} < \pi(n) < 6 \cdot \frac{n}{\log n}$$

□

Theorem 1.22 (DIRICHLET'S APPROXIMATION THEOREM) *Let A be a real number and n a natural number. Then, there exists an integer b, such that $0 < b \leq n$, and an integer c, for which the following holds*

$$|Ab - c| < \frac{1}{n}.$$

Proof Let $\{Ai\} = Ai - \lfloor Ai \rfloor$, for $i = 0, 1, 2, \ldots, n$. It is clear that $0 \leq \{Ai\} < 1$. We now construct the intervals

$$\left[\frac{x}{n}, \frac{x+1}{n}\right),$$

where $0 \leq x < n$.
Since there are $n + 1$ real numbers $\{Ai\}$, such that $0 \leq \{Ai\} < 1$, by the Pigeonhole Principle it follows that at least one of the intervals $[x/n, (x + 1)/n)$ will contain two of these numbers.
Let us suppose that

$$\{Ak\}, \{Al\} \in \left[\frac{x}{n}, \frac{x+1}{n}\right),$$

for some $0 \leq x < n$.
Therefore

$$|\{Ak\} - \{Al\}| < \frac{1}{n}$$

or

$$|Ak - \lfloor Ak \rfloor - (Al - \lfloor Al \rfloor)| < \frac{1}{n}$$

or

$$|A(k - l) - (\lfloor Ak \rfloor - \lfloor Al \rfloor)| < \frac{1}{n}.$$

Thus, we distinguish the following cases:

- If $k - l > 0$, then we set $b = k - l$ and $c = \lfloor Ak \rfloor - \lfloor Al \rfloor$.
- If $k - l < 0$, then we set $b = l - k$ and $c = \lfloor Al \rfloor - \lfloor Ak \rfloor$.

Hence, we obtain

$$|Ab - c| < \frac{1}{n} .$$

□

Corollary 1.23 *Let A be a real number and n a natural number. Then, there exists an integer b, such that* $0 < b \leq n$, *and an integer c relatively prime to b, for which it holds*

$$\left|A - \frac{c}{b}\right| < \frac{1}{b^2} .$$

Proof By Dirichlet's Approximation Theorem, we have

$$|Ab - c| < \frac{1}{n} .$$

Thus, since b is a positive integer, we can write

$$\frac{|Ab - c|}{b} < \frac{1}{nb} \leq \frac{1}{b^2}$$

or

$$\left|A - \frac{c}{b}\right| < \frac{1}{b^2} .$$

□

The following theorem as well as other related theorems can be found in [9, 23].

Theorem 1.24 (Siegel- Walfisz Theorem)
Let D be a positive constant. Then there exists a positive constant $C(D)$ *such that the following holds: Assume that r is a real number and a, q are integers such that* $(a, q) = 1$ *with* $q \leq \log^D r$. *Then*

$$\sum_{\substack{n \leq r \\ n \equiv a (\bmod q)}} \Lambda(n) = \frac{r}{\phi(q)} + O\left(r \exp\left(-C(D)\sqrt{\log r}\right)\right) ,$$

where $\Lambda(n)$ *denotes the Von Mangoldt function and* $\phi(n)$ *the Euler totient function.*

2 The Circle Method

The Circle Method was introduced for the first time in a paper by Hardy and Ramanujan [21] concerning partitions. Moreover, Hardy and Littlewood developed that method so that it could be used to connect exponential sums with general problems of additive number theory.[3] For recent developments and generalizations of the Hardy-Littlewood method to additive number theory, the interested reader is referred to the paper of Green [17].
A characteristic problem to which the Circle Method finds an application is the following:

Problem 2.1 *Let S be a subset of $\mathbb{N}$ and $k \in \mathbb{N}$. Determine*

$$\{s_1 + s_2 + \cdots + s_k \mid s_1, s_2, \ldots, s_k \in S\} \cap \mathbb{N}.$$

In other words, determine which natural numbers can be represented as the sum of k elements of the set S and in how many ways.

Remark 2.2 If we set $S = \mathbb{P}$, where $\mathbb{P}$ denotes the set of all prime numbers, then

1. For $k = 2$, the statement of Problem 2.1 becomes:
   Determine the set

$$E = \{p_1 + p_2 \mid p_1, p_2 \in \mathbb{P}\} \cap \mathbb{N}.$$

   The Goldbach conjecture states that the set E is the set of all positive even integers.
2. For $k = 3$, the statement of Problem 2.1 becomes:
   Determine the set

$$O = \{p_1 + p_2 + p_3 \mid p_1, p_2, p_3 \in \mathbb{P}\} \cap \mathbb{N}.$$

In 1937, I. M. Vinogradov [66, 67] proved that every large enough odd positive integer is included in the set O.
Generally, the starting point of the Circle Method is to consider a generating function of the form:

$$F_S(x) = \sum_{s \in S} x^s.$$

[3]Hardy and Littlewood in a paper published in 1923 have used the Circle Method to prove that on assumption of a modified form of the Riemann Hypothesis there exists a natural number N, such that every odd integer $n \geq N$ can be expressed as the sum of three prime numbers.

Questions of convergence may be avoided if S is a finite set, which we shall assume in the following. We write

$$F_S(x)^k = \sum_{n=1}^{+\infty} R(n, k, S)x^n .$$

It can be proved that the coefficient $R(n, k, S)$ is equal to the number of ways that n can be represented as the sum of k elements of the set S.
Moreover, it follows from Cauchy's formula that

$$R(n, k, S) = \frac{1}{2\pi i} \int_C \frac{F_S(z)^k}{z^{n+1}} dz , \tag{1}$$

where C is the unit circle oriented counterclockwise.
However, if we substitute $x = e^{2\pi i u}$ and

$$f_S(u) = F_S(x),$$

we obtain

$$R(n, k, S) = \int_0^1 f_S(u)^k e^{-2\pi i n u} du .$$

In addition, for every natural number $n \leq N$, it holds

$$R(n, k, S) = R_N(n, k, S) = \int_0^1 f_N(x)^k e^{-2\pi i n x} dx ,$$

where $R_N(n, k, S)$ is equal to the number of ways that n can be represented as the sum of k elements of the set S, where each element is at most N.
The key feature of the Circle Method is to split C into two disjoint pieces, generally referred to as the *Major* and *Minor arcs* $\mathfrak{M}$ and $\mathfrak{m}$, respectively. Therefore, we obtain

$$R(n, k, S) = R_N(n, k, S) = \int_{\mathfrak{M}} f_N(x)^k e^{-2\pi i n x} dx + \int_{\mathfrak{m}} f_N(x)^k e^{-2\pi i n x} dx$$

or equivalently

$$R(n, k, S) = R_N(n, k, S) = \int_{\mathfrak{M}} f_N(x)^k e(-nx) dx + \int_{\mathfrak{m}} f_N(x)^k e(-nx) dx .$$

The basic idea behind the choice of the Major and Minor arcs is the following: The Major arcs are constructed in such a way, so that the function in the integral

$$\int_{\mathfrak{M}}$$

can be evaluated asymptotically and that the contribution of the Minor arcs is of lower order.

3 Proof of Vinogradov's Theorem

The purpose of this section is to present R. C. Vaughan's proof of Vinogradov's theorem.

Theorem 3.1 (VINOGRADOV'S THEOREM)
There exists a natural number N, such that every odd positive integer n, with $n \geq N$, can be represented as the sum of three prime numbers.

Before we define the appropriate function f and construct the relevant Major and Minor arcs, in order to apply the Circle Method, we observe that

$$\begin{aligned} R(n, 3, \mathbb{P}) &= \sum_{n=p_1+p_2+p_3} 1 \\ &> \sum_{n=p_1+p_2+p_3} \frac{\log p_1 \cdot \log p_2 \cdot \log p_3}{\log^3 (p_1+p_2+p_3)} \\ &= \sum_{n=p_1+p_2+p_3} \frac{\log p_1 \cdot \log p_2 \cdot \log p_3}{\log^3 n} \end{aligned}$$

or equivalently

$$R(n, 3, \mathbb{P}) > \frac{1}{\log^3 n} \sum_{n=p_1+p_2+p_3} \log p_1 \cdot \log p_2 \cdot \log p_3 \,, \tag{a}$$

where $\mathbb{P}$ denotes the set of all prime numbers and consequently p_1, p_2, p_3 are prime numbers.
Therefore, instead of working with the sum

$$\sum_{n=p_1+p_2+p_3} 1$$

we shall work with the sum

$$\sum_{n=p_1+p_2+p_3} \log p_1 \cdot \log p_2 \cdot \log p_3$$

More specifically, Vinogradov succeeded in proving that

$$\sum_{n=p_1+p_2+p_3} \log p_1 \cdot \log p_2 \cdot \log p_3 \gg n^2 .$$

Thus, by (a), we obtain

$$R(n, 3, \mathbb{P}) \gg \frac{n^2}{\log^3 n} ,$$

from which it is obvious that there exists a natural number N, such that every $n \geq N$, can be represented as the sum of three prime numbers.
Let us now proceed to the details of the proof of Vinogradov's Theorem by the use of the Circle Method.
Let

$$f(x) = \sum_{p \leq N} \log p \cdot e(xp)$$

and

$$f_r(x) = \sum_{p \leq r} \log p \cdot e(xp) ,$$

where p is a prime number and x, r are real numbers.
In addition, let

$$\overline{R}_N(m, k) = \sum_{\substack{m=p_1+p_2+\cdots+p_k \\ p_i \leq N}} \log p_1 \cdot \log p_2 \cdots \log p_k ,$$

where $p_1, p_2, \ldots, p_k$ are prime numbers.
Then, it follows that

$$\overline{R}_N(m, k) = \int_0^1 f^k(x) e(-mx) dx$$

and in our case, for $k = 3$, one has

$$\overline{R}_N(m, 3) = \sum_{\substack{m=p_1+p_2+p_3 \\ p_i \leq N}} \log p_1 \cdot \log p_2 \cdot \log p_3 .$$

We shall now construct the Major and Minor arcs. As we briefly mentioned in the section concerning the Circle Method, we have to split the unit circle C into two disjoint pieces (equivalently we can split the interval $[0, 1]$ into two disjoint pieces).
Since in this problem we are going to make use of the Siegel-Walfisz Theorem 1.24, it is evident that we must first consider a positive constant D and set

$$L = \log^D N .$$

More specifically, we consider D, such that $D > 10$.
We define the Major arcs as follows:

$$\mathfrak{M} = \bigcup_{\substack{1 \le q \le L \\ (a,q)=1}} \mathfrak{M}_{(a,q)} ,$$

where

$$\mathfrak{M}_{(a,q)} = \left\{ x \in \left[\frac{L}{N}, 1 + \frac{L}{N} \right] : \left| x - \frac{a}{q} \right| \le \frac{L}{N} \right\}$$

and $a \in \{1, 2, \dots, q\}$.
At this point, we shall prove a useful lemma.

Lemma 3.2 *Let a, q be positive integers such that $1 \le a \le q$, $1 \le q \le L$ and $(a, q) = 1$. Then, for all sufficiently large N, the Major arcs $\mathfrak{M}$ can be expressed as a disjoint union of $\mathfrak{M}_{(a,q)}$.*

Proof Let us suppose that there exists $x \in \mathfrak{M}_{(a_1,q_1)} \cap \mathfrak{M}_{(a_2,q_2)}$, with

$$\left| \frac{a_1}{q_1} - \frac{a_2}{q_2} \right| > 0 .$$

Then, it is evident that

$$|a_1 q_2 - a_2 q_1| > 0$$

or

$$|a_1 q_2 - a_2 q_1| \ge 1 .$$

However,

$$\begin{aligned}\frac{2L}{N} &\geq \left|x - \frac{a_2}{q_2}\right| + \left|\frac{a_1}{q_1} - x\right| \geq \left|\frac{a_1}{q_1} - \frac{a_2}{q_2}\right| \\ &= \left|\frac{a_1 q_2 - a_2 q_1}{q_1 q_2}\right| \geq \frac{1}{q_1 q_2} \\ &\geq \frac{1}{L^2}.\end{aligned}$$

Therefore, we have

$$2L^3 \geq N.$$

But, by the definition of L we obtain

$$2\log^{3D} N \geq N,$$

which is not true for large values of N. Hence, we have arrived to a contradiction. This completes the proof of the lemma. □

We define now the Minor arcs $\mathfrak{m}$ as follows:

$$\mathfrak{m} = \left[\frac{L}{N}, 1 + \frac{L}{N}\right] \setminus \mathfrak{M}.$$

3.1 *The Contribution of the Major Arcs*

In this section, we shall investigate the contribution of the Major arcs by proving two basic theorems. The first one provides an approximation of $f(x)$ for $x \in \mathfrak{M}_{(a,q)}$ and the second one provides an approximation of the integral

$$\int_{\mathfrak{M}} f^3(x) e(-xN) dx.$$

Theorem 3.3 *Let $x \in \mathfrak{M}_{(a,q)}$. Then there exists a positive constant C, such that*

$$f(x) - \frac{\mu(q)}{\phi(q)} \sum_{n=1}^{N} e\left(\left(x - \frac{a}{q}\right) n\right) \ll N \exp\left(-C\sqrt{\log N}\right).$$

Proof Let r be a real number, such that $r \in [1, N]$. Then, it holds

$$f_r\left(\frac{a}{q}\right) = \sum_{p \leq r} \log p \cdot e\left(\frac{a}{q} p\right).$$

But, it is clear that

$$p \equiv t \pmod{q},$$

for some integer t with $1 \leq t \leq q$.
Therefore, we can write

$$\begin{aligned}
f_r\left(\frac{a}{q}\right) &= \sum_{t=1}^{q} \sum_{\substack{p \equiv t (\bmod q) \\ p \leq r}} \log p \cdot e\left(\frac{a}{q} p\right) \\
&= \sum_{t=1}^{q} \sum_{\substack{p \equiv t (\bmod q) \\ p \leq r}} \log p \cdot e\left(\frac{a}{q} t\right) \\
&= \sum_{t=1}^{q} \left(e\left(\frac{a}{q} t\right) \sum_{\substack{p \equiv t (\bmod q) \\ p \leq r}} \log p \right) \\
&= \sum_{\substack{t=1 \\ (t,q)=1}}^{q} e\left(\frac{a}{q} t\right) \sum_{\substack{p \equiv t (\bmod q) \\ p \leq r}} \log p + \sum_{\substack{t=1 \\ (t,q)>1}}^{q} e\left(\frac{a}{q} t\right) \sum_{\substack{p \equiv t (\bmod q) \\ p \leq r}} \log p .
\end{aligned}$$

Hence,

$$\begin{aligned}
&\left| f_r\left(\frac{a}{q}\right) - \frac{r}{\phi(q)} \sum_{\substack{t=1 \\ (t,q)=1}}^{q} e\left(\frac{at}{q}\right) \right| \\
&= \left| \sum_{\substack{t=1 \\ (t,q)=1}}^{q} e\left(\frac{at}{q}\right) \left(\sum_{\substack{p \leq r \\ p \equiv t (\bmod q)}} \log p - \frac{r}{\phi(q)} \right) + \sum_{\substack{t=1 \\ (t,q)>1}}^{q} e\left(\frac{at}{q}\right) \sum_{\substack{p \leq r \\ p \equiv t (\bmod q)}} \log p \right| \\
&\leq \sum_{\substack{t=1 \\ (t,q)=1}}^{q} \left| e\left(\frac{at}{q}\right) \right| \left| \sum_{\substack{p \leq r \\ p \equiv t (\bmod q)}} \log p - \frac{r}{\phi(q)} \right| + \sum_{\substack{t=1 \\ (t,q)>1}}^{q} \left| e\left(\frac{at}{q}\right) \right| \left(\sum_{\substack{p \leq r \\ p \equiv t (\bmod q)}} \log p \right) .
\end{aligned}$$

Therefore, we get

$$\left| f_r\left(\frac{a}{q}\right) - \frac{r}{\phi(q)} \sum_{\substack{t=1 \\ (t,q)=1}}^{q} e\left(\frac{at}{q}\right) \right|$$

$$\leq \sum_{\substack{t=1 \\ (t,q)=1}}^{q} \left| \sum_{\substack{p \leq r \\ p \equiv t (\bmod q)}} \log p - \frac{r}{\phi(q)} \right| + \sum_{\substack{t=1 \\ (t,q)>1}}^{q} \left(\sum_{\substack{p \leq r \\ p \equiv t (\bmod q)}} \log p \right). \tag{1}$$

By Theorem 1.24, we have

$$\sum_{\substack{p \leq r \\ p \equiv t (\bmod q)}} \log p - \frac{r}{\phi(q)} = O\left(r \exp\left(-C_D \sqrt{\log r} \right) \right).$$

Thus,

$$\left| \sum_{\substack{p \leq r \\ p \equiv t (\bmod q)}} \log p - \frac{r}{\phi(q)} \right| \ll r \exp\left(-C_D \sqrt{\log r} \right). \tag{2}$$

Since, $1 \leq r \leq N$, we get

$$\begin{aligned} r \exp\left(-C_D \sqrt{\log r} \right) &= r e^{-C_D \sqrt{\log r}} = e^{\log r - C_D \sqrt{\log r}} \\ &= e^{\sqrt{\log r}(\sqrt{\log r} - C_D)} \leq e^{\sqrt{\log N}(\sqrt{\log N} - C_D)} \\ &= e^{\log N - C_D \sqrt{\log N}} = N \exp\left(-C_D \sqrt{\log N} \right). \end{aligned}$$

Therefore, by (2), we obtain

$$\left| \sum_{\substack{p \leq r \\ p \equiv t (\bmod q)}} \log p - \frac{r}{\phi(q)} \right| \ll N \exp\left(-C_D \sqrt{\log N} \right). \tag{3}$$

However, by the definition of the Euler ϕ function, we know that

$$\phi(q) = \sum_{\substack{t=1 \\ (t,q)=1}}^{q} 1 \,,$$

and thus, by (3), it is clear that

$$\sum_{\substack{t=1 \\ (t,q)=1}}^{q} \left| \sum_{\substack{p \le r \\ p \equiv t (\bmod q)}} \log p - \frac{r}{\phi(q)} \right| \ll \sum_{\substack{t=1 \\ (t,q)=1}}^{q} N \exp\left(-C_D \sqrt{\log N}\right)$$

$$= N \exp\left(-C_D \sqrt{\log N}\right) \cdot \sum_{\substack{t=1 \\ (t,q)=1}}^{q} 1 \,,$$

or

$$\sum_{\substack{t=1 \\ (t,q)=1}}^{q} \left| \sum_{\substack{p \le r \\ p \equiv t (\bmod q)}} \log p - \frac{r}{\phi(q)} \right| \ll N \exp\left(-C_D \sqrt{\log N}\right) \cdot \phi(q) \,. \tag{4}$$

We also have

$$\sum_{\substack{t=1 \\ (t,q)>1}}^{q} \left(\sum_{\substack{p \le r \\ p \equiv t (\bmod q)}} \log p \right) \ll \sum_{p|q} \log p \,. \tag{5}$$

Hence, by (1), (4) and (5), it follows that

$$\left| f_r\left(\frac{a}{q}\right) - \frac{r}{\phi(q)} \sum_{\substack{t=1 \\ (t,q)=1}}^{q} e\left(\frac{at}{q}\right) \right| \ll \phi(q) N \exp\left(-C_D \sqrt{\log N}\right) + \sum_{p|q} \log p \,.$$

Since $q \le L = \log^D N$, it is evident that

$$\phi(q) = \sum_{\substack{t=1 \\ (t,q)=1}}^{q} 1 \le L \,.$$

Therefore,

$$\left| f_r\left(\frac{a}{q}\right) - \frac{r}{\phi(q)} \sum_{\substack{t=1 \\ (t,q)=1}}^{q} e\left(\frac{at}{q}\right) \right| \ll LN \exp\left(-C_D\sqrt{\log N}\right) + \sum_{p|q} \log p$$

$$\leq LN \exp\left(-C_D\sqrt{\log N}\right) + \log N \,.$$

By the above relation, it is clear that

$$\left| f_r\left(\frac{a}{q}\right) - \frac{r}{\phi(q)} \sum_{\substack{t=1 \\ (t,q)=1}}^{q} e\left(\frac{at}{q}\right) \right| \ll N \exp\left(-C\sqrt{\log N}\right) \,, \tag{6}$$

for any positive constant $C < C_D$.
However, by the definition of the Ramanujan sum, we have

$$c_q(a) = \sum_{\substack{t=1 \\ (t,q)=1}}^{q} e\left(\frac{at}{q}\right)$$

and thus (6) takes the form

$$\left| f_r\left(\frac{a}{q}\right) - \frac{rc_q(a)}{\phi(q)} \right| \ll N \exp\left(-C\sqrt{\log N}\right) \,, \tag{7}$$

But, by the hypothesis of the theorem and Lemma 1.16, it follows that in this case

$$c_q(a) = \mu(q) \,.$$

Therefore, (7) is equivalent to

$$\left| f_r\left(\frac{a}{q}\right) - r\frac{\mu(q)}{\phi(q)} \right| \ll N \exp\left(-C\sqrt{\log N}\right) \,, \tag{8}$$

Now, let

$$E_d := (\pi(d) - \pi(d-1))\, e\left(\frac{ad}{q}\right) \log d - \frac{\mu(q)}{\phi(q)} \,,$$

where $\pi(x)$ denotes the prime counting function[4].

---

[4]It is evident that if d is a prime number, then $\pi(d) - \pi(d-1) = 1$, and thus,

We have

$$\left|f(x) - \frac{\mu(q)}{\phi(q)} \sum_{d=1}^{N} e(wd)\right| = \left|\sum_{d=1}^{N} E_d e(wd)\right| ,$$

where $w = x - a/q$.
But, it is clear that

$$e(wd) = e(wN) - \int_d^N \frac{d}{dy}\, e(wy)dy\,.$$

Hence, we obtain

$$\begin{aligned}\left|f(x) - \frac{\mu(q)}{\phi(q)} \sum_{d=1}^{N} e(wd)\right| &= \left|e(wN) \sum_{d=1}^{N} E_d - \sum_{d=1}^{N} \left(E_d \int_d^N \frac{d}{dy}\, e(wy)dy\right)\right| \\ &\le \left|e(wN) \sum_{d=1}^{N} E_d\right| + \left|\sum_{d=1}^{N} \left(E_d \int_d^N \frac{d}{dy}\, e(wy)dy\right)\right| \\ &= \left|\sum_{d=1}^{N} E_d\right| + \left|\int_0^N \left(\frac{d}{dy}\, e(wy) \sum_{d=1}^{y} E_d\right) dy\right| \\ &\le \left|\sum_{d=1}^{N} E_d\right| + \int_0^N |2\pi i w e(wy)| \left|\sum_{d=1}^{y} E_d\right| dy\,.\end{aligned}$$

However,

$$\sum_{d=1}^{r} E_d = f_r\left(\frac{a}{q}\right) - r\frac{\mu(q)}{\phi(q)}\,.$$

Thus, by (8), we get that each one of

$$\left|\sum_{d=1}^{N} E_d\right|, \left|\sum_{d=1}^{y} E_d\right| \ll N \exp\left(-C\sqrt{\log N}\right)$$

and therefore, since

$$|w| \le \frac{L}{N}\,,$$

(Footnote 4 continued)

$$E_d = e(ad/q)\log d - \mu(q)/\phi(q)\,.$$

On the other hand, if d is a composite number $\pi(d) = \pi(d-1)$, which yields $E_d = -\mu(q)/\phi(q)$.

we obtain

$$\begin{aligned}\left|f(x)-\frac{\mu(q)}{\phi(q)}\sum_{d=1}^{N}e(wd)\right| &\ll N\exp\left(-C\sqrt{\log N}\right)+(2\pi\,|w|\,N)\,N\exp\left(-C\sqrt{\log N}\right)\\ &\leq N\exp\left(-C\sqrt{\log N}\right)+\left(2\pi\frac{L}{N}\right)N\exp\left(-C\sqrt{\log N}\right)\\ &=(1+2\pi L)\,N\exp\left(-C\sqrt{\log N}\right)\\ &\ll N\exp\left(-C'\sqrt{\log N}\right),\end{aligned}$$

for any constant $C' < C$.
This completes the proof of Theorem 3.3. □

Theorem 3.4 *Let*

$$G(N) := \sum_{q=1}^{+\infty}\frac{\mu(q)c_q(N)}{\phi(q)^3},$$

where $c_q(N)$ stands for the Ramanujan sum.
Then,

$$\int_{\mathfrak{M}} f^3(x)e(-xN)dx - \frac{N^2}{2}G(N) \ll N^2\log^{-D/2}N.$$

Proof Let $w = x - a/q$. Then,

$$\left|f^3(x)-\frac{\mu(q)^3}{\phi(q)^3}\left(\sum_{d=1}^{N}e(wd)\right)^3\right|$$

$$=\left|f(x)-\frac{\mu(q)}{\phi(q)}\sum_{d=1}^{N}e(wd)\right|\cdot\left|f^2(x)-f(x)\frac{\mu(q)}{\phi(q)}\sum_{d=1}^{N}e(wd)+\frac{\mu(q)^2}{\phi(q)^2}\left(\sum_{d=1}^{N}e(wd)\right)^2\right|$$

$$\leq\left|f(x)-\frac{\mu(q)}{\phi(q)}\sum_{d=1}^{N}e(wd)\right|\left(\left|f^2(x)\right|+|f(x)|\left|\frac{\mu(q)}{\phi(q)}\sum_{d=1}^{N}e(wd)\right|+\left|\frac{\mu(q)^2}{\phi(q)^2}\left(\sum_{d=1}^{N}e(wd)\right)^2\right|\right).$$

However, it is evident that

$$|f(x)| \leq \pi(N)\log N$$

and by Chebyshev's inequality (Theorem 1.21), it follows that

$$|f(x)| \ll N .$$

In addition, it is clear that

$$\left| \frac{\mu(q)}{\phi(q)} \sum_{d=1}^{N} e(wd) \right| \ll N .$$

Moreover, since all the possible values of $\mu(q)$ are -1, 0 and 1, it is obvious that $\mu(q)^3 = \mu(q)$. Therefore, by all the above, we obtain

$$\left| f^3(x) - \frac{\mu(q)}{\phi(q)^3} \left(\sum_{d=1}^{N} e(wd) \right)^3 \right| \ll 3N^2 \left| f(x) - \frac{\mu(q)}{\phi(q)} \sum_{d=1}^{N} e(wd) \right| . \quad (1)$$

But, by the previous theorem, we know that for $x \in \mathfrak{M}_{(a,q)}$, it holds

$$f(x) - \frac{\mu(q)}{\phi(q)} \sum_{d=1}^{N} e(wd) \ll N \exp\left(-C\sqrt{\log N}\right) .$$

Thus, by (1) we get

$$\left| f^3(x) - \frac{\mu(q)}{\phi(q)^3} \left(\sum_{d=1}^{N} e(wd) \right)^3 \right| \ll N^3 \exp\left(-C\sqrt{\log N}\right) .$$

Since,

$$\mathfrak{M} = \bigcup_{\substack{1 \le q \le L \\ (a,q)=1}} \mathfrak{M}_{(a,q)} ,$$

in order to obtain the integral over the Major arcs $\mathfrak{M}$, we must integrate over $\mathfrak{M}_{(a,q)}$ and sum over all q, $1 \le q \le L$ and all a, $1 \le a \le q$ with $(a, q) = 1$. However,

$$\sum_{q=1}^{L} \sum_{\substack{a=1 \\ (a,q)=1}}^{q} \int_{\mathfrak{M}_{(a,q)}} \left(f^3(x) - \frac{\mu(q)}{\phi(q)^3} \left(\sum_{d=1}^{N} e(wd) \right)^3 \right) e(-xN)\, dx$$

$$\leq \sum_{q=1}^{L} \sum_{\substack{a=1 \\ (a,q)=1}}^{q} \int_{\mathfrak{M}_{(a,q)}} \left| f^3(x) - \frac{\mu(q)}{\phi(q)^3} \left(\sum_{d=1}^{N} e(wd) \right)^3 \right| dx$$

$$\ll \sum_{q=1}^{L} \sum_{\substack{a=1 \\ (a,q)=1}}^{q} \int_{\mathfrak{M}_{(a,q)}} N^3 \exp\left(-C\sqrt{\log N}\right) dx$$

$$\leq L^2 \cdot 2\,\frac{L}{N} \cdot N^3 \exp\left(-C\sqrt{\log N}\right), \text{ since } |\mathfrak{M}_{(a,q)}| = 2\,\frac{L}{N}$$

$$= 2\,\frac{L^{3+1/2}}{L^{1/2}} \cdot N^2 \exp\left(-C\sqrt{\log N}\right).$$

But, by the definition of L, we have

$$L^g = \log^{gD} N = \exp\left(\log\left(gD \log N\right)\right) \ll \exp\left(C'\sqrt{\log N}\right),$$

for any positive constant C' and $g \geq 1$.
Thus, for $g = 3 + 1/2$ and $C' = C$, we obtain

$$S := \sum_{q=1}^{L} \sum_{\substack{a=1 \\ (a,q)=1}}^{q} \int_{\mathfrak{M}_{(a,q)}} \left(f^3(x) - \frac{\mu(q)}{\phi(q)^3} \left(\sum_{d=1}^{N} e(wd) \right)^3 \right) e(-xN)dx$$

$$\ll \frac{N^2 \exp\left(C\sqrt{\log N}\right)}{L^{1/2}} \cdot \exp\left(-C\sqrt{\log N}\right)$$

$$= \frac{N^2}{L^{1/2}}.$$

Moreover, we have

$$S = \int_{\mathfrak{M}} f^3(x)e(-xN)dx - \sum_{q=1}^{L} \sum_{\substack{a=1 \\ (a,q)=1}}^{q} \int_{\mathfrak{M}_{(a,q)}} \frac{\mu(q)}{\phi(q)^3} \left(\sum_{d=1}^{N} e(wd) \right)^3 e(-xN)dx$$

$$= \int_{\mathfrak{M}} f^3(x)e(-xN)dx$$

$$- \sum_{q=1}^{L} \sum_{\substack{a=1 \\ (a,q)=1}}^{q} \int_{\mathfrak{M}_{(a,q)}} \frac{\mu(q)}{\phi(q)^3} \left(\sum_{d=1}^{N} e(wd) \right)^3 e\left(-\left(x - \frac{a}{q}\right)N\right) \cdot e\left(-\frac{a}{q}N\right) dx$$

$$= \int_{\mathfrak{M}} f^3(x)e(-xN)dx - \sum_{q=1}^{L} \sum_{\substack{a=1 \\ (a,q)=1}}^{q} e\left(-\frac{aN}{q}\right) \int_{\mathfrak{M}_{(a,q)}} \frac{\mu(q)}{\phi(q)^3} \left(\sum_{d=1}^{N} e(wd) \right)^3 e(-wN)dw$$

(note that $w = x - a/q$) and since we proved that

$$S \ll \frac{N^2}{L^{1/2}},$$

it is evident that

$$\int_{\mathfrak{M}} f^3(x)e(-xN)dx - \sum_{q=1}^{L} \sum_{\substack{a=1 \\ (a,q)=1}}^{q} e\left(-\frac{aN}{q}\right) \frac{\mu(q)}{\phi(q)^3} \int_{-\frac{L}{N}}^{\frac{L}{N}} \left(\sum_{d=1}^{N} e(wd)\right)^3 e(-wN)dw \ll \frac{N^2}{L^{1/2}}. \tag{2}$$

Therefore, by (2) we see that we must also determine a bound for the integral

$$I = \int_{-\frac{L}{N}}^{\frac{L}{N}} \left(\sum_{d=1}^{N} e(wd)\right)^3 e(-wN)dw .$$

However, we observe that

$$\begin{aligned} I' &:= \int_{-\frac{1}{2}}^{\frac{1}{2}} \left(\sum_{d=1}^{N} e(wd)\right)^3 e(-wN)dw \\ &= \sum_{\substack{d_1+d_2+d_3=N \\ d_i \geq 1}} 1 \\ &= \frac{(N-1)(N-2)}{2} \end{aligned} \tag{3}$$

and therefore

$$\left| I' - \frac{N^2}{2} \right| \leq 2N .$$

Since we know the exact value of I', we shall try to correlate the integral I with the integral I'. Let

$$h(w) = \left(\sum_{d=1}^{N} e(wd)\right)^3 e(-wN) .$$

Therefore, we have

$$
\begin{aligned}
|I' - I| &= \left| \int_{-\frac{1}{2}}^{\frac{1}{2}} h(w)dw - \left(\int_{-\frac{L}{N}}^{-\frac{1}{2}} h(w)dw + \int_{-\frac{1}{2}}^{\frac{1}{2}} h(w)dw + \int_{\frac{1}{2}}^{\frac{L}{N}} h(w)dw \right) \right| \\
&= \left| \int_{-\frac{L}{N}}^{-\frac{1}{2}} h(w)dw + \int_{\frac{1}{2}}^{\frac{L}{N}} h(w)dw \right| .
\end{aligned}
$$

If we substitute w with $-w$, we get

$$
\begin{aligned}
|I' - I| &= \left| - \int_{\frac{L}{N}}^{\frac{1}{2}} h(-w)dw + \int_{\frac{1}{2}}^{\frac{L}{N}} h(w)dw \right| \\
&= \left| - \int_{\frac{L}{N}}^{\frac{1}{2}} h(-w)dw - \int_{\frac{L}{N}}^{\frac{1}{2}} h(w)dw \right| \\
&\leq \int_{\frac{L}{N}}^{\frac{1}{2}} \left| \sum_{d=1}^{N} e(-wd) \right|^3 dw + \int_{\frac{L}{N}}^{\frac{1}{2}} \left| \sum_{d=1}^{N} e(wd) \right|^3 dw \\
&\leq 2 \int_{\frac{L}{N}}^{\frac{1}{2}} \left| \sum_{d=1}^{N} e(-wd) \right|^3 dw .
\end{aligned}
$$

But, by Lemma 1.17, it follows that

$$
\left| \sum_{d=1}^{N} e(-wd) \right| \leq \min \left\{ \frac{1}{[-w]}, N \right\} = \frac{1}{[w]} .
$$

Thus,

$$
|I' - I| \leq 2 \int_{\frac{L}{N}}^{\frac{1}{2}} \frac{1}{[w]^3} dw = 2 \int_{\frac{L}{N}}^{\frac{1}{2}} \frac{1}{w^3} dw = \frac{N^2}{L^2} - 4 < \left(\frac{N}{L} \right)^2 .
$$

It follows that

$$
\left| \sum_{\substack{a=1 \\ (a,q)=1}}^{q} \frac{\mu(q)}{\phi(q)^3} e\left(-\frac{aN}{q} \right) \right| = \left| \frac{\mu(q)}{\phi(q)^3} \sum_{\substack{a=1 \\ (a,q)=1}}^{q} e\left(-\frac{aN}{q} \right) \right|
$$

$$\leq \frac{|\mu(q)|}{\phi(q)^3} \sum_{\substack{a=1 \\ (a,q)=1}}^{q} 1$$

$$\leq \frac{1}{\phi(q)^3}\phi(q) = \frac{1}{\phi(q)^2} .$$

But, it can be shown that if β is a positive real number, then

$$\lim_{n\to\infty} \frac{n^{1-\beta}}{\phi(n)} = 0 .$$

Thus, it is evident that for $\beta = 1/4$ there exists sufficiently large N, such that

$$\frac{N^{1-1/4}}{\phi(N)} < 1$$

or

$$N^{3/4} < \phi(N) .$$

Therefore, it is clear that

$$\frac{1}{\phi(q)^2} \ll \frac{1}{q^{3/2}} .$$

Hence, we have with

$$G(N) := \sum_{q=1}^{+\infty} \frac{\mu(q)c_q(N)}{\phi(q)^3} ,$$

that

$$\left| G(N) - \sum_{q=1}^{L} \sum_{\substack{a=1 \\ (a,q)=1}}^{q} \frac{\mu(q)}{\phi(q)^3} e\left(-\frac{aN}{q}\right) \right| \ll \sum_{q=L+1}^{+\infty} \frac{1}{q^{3/2}}$$

$$\leq \int_{L+1}^{+\infty} \frac{1}{x^{3/2}} dx$$

$$= \frac{2}{(L+1)^{1/2}} .$$

Thus,

$$\left| G(N) - \sum_{q=1}^{L} \sum_{\substack{a=1 \\ (a,q)=1}}^{q} \frac{\mu(q)}{\phi(q)^3} e\left(-\frac{aN}{q}\right) \right| \ll \frac{1}{L^{1/2}} . \tag{4}$$

By (2) and (3) we obtain

$$\int_{\mathfrak{M}} f^3(x)e(-xN)dx - \left(\sum_{q=1}^{L} \sum_{\substack{a=1 \\ (a,q)=1}}^{q} e\left(-\frac{aN}{q}\right) \frac{\mu(q)}{\phi(q)^3} \right) \cdot \frac{N^2}{2} \ll \frac{N^2}{L^{1/2}} + \frac{N^2}{L^2} + N$$

$$= (1 + L^{-3/2}) \frac{N^2}{L^{1/2}}$$

$$\ll \frac{N^2}{L^{1/2}} .$$

By the above relation and (3), we get

$$\int_{\mathfrak{M}} f^3(x)e(-xN)dx - G(N)\frac{N^2}{2} \ll \frac{N^2}{L^{1/2}} = N^2 \log^{-D/2} N .$$

This completes the proof of Theorem 3.4. □

3.2 *The Contribution of the Minor Arcs*

In this section, we shall investigate the contribution of the Minor arcs $\mathfrak{m}$. Our ultimate goal is to prove that

$$\int_{\mathfrak{m}} f^3(x)e(-xN)dx \ll \frac{N^2}{\log^c N} ,$$

for any positive constant c with $c \leq D/2 - 5$ and D as in Theorem 1.24 where

$$f(x) = \sum_{p \leq N} \log p \cdot e(xp)$$

We need to prove some more theorems and lemmas.
The following lemma is presented without a proof, since it is a classical result in approximation theory and analytic number theory.

Lemma 3.5 *For any real numbers x, r_1, r_2, with $r_1, r_2 \geq 1$, and integers q, a, such that*

$$\left|x - \frac{a}{q}\right| \leq \frac{1}{q^2}\,,\ (a,q) = 1$$

and $q \geq 1$, it holds

$$\sum_{n<r_1} \min\left\{\frac{1}{[xn]}, \frac{r_1 r_2}{n}\right\} \ll \left(\frac{r_1 r_2}{q} + r_1 + q\right) \log(2r_1 q)\,. \tag{1}$$

The following result is a special case of an identity due to R. C. Vaughan [64].

Lemma 3.6 *Let r be a real number, such that $1 \leq r \leq \sqrt{N}$.* [5]
Then,

$$\sum_{r<k\leq N} \Lambda(k)e(xk) = \sum_{d\leq r}\sum_{m\leq \frac{N}{d}} \log m \cdot \mu(d)e(xdm)$$

$$-\sum_{r<d\leq N}\ \sum_{r<m\leq \frac{N}{d}}\ \sum_{\substack{q|d \\ q\leq r}} \mu(q)\Lambda(m)e(xdm)$$

$$-\sum_{d\leq r^2}\sum_{m\leq \frac{N}{d}}\sum_{\substack{q|d \\ q\leq r \\ d\leq rq}} \mu(q)\Lambda\left(\frac{d}{q}\right)e(xdm)\,, \tag{L1}$$

where $\mu(n)$ and $\Lambda(n)$ denote the Möbius and the von Mangoldt function, respectively.

Proof Throughout the proof, we assume that $Re\{s\} > 1$. All the Dirichlet series considered in the proof will be absolutely convergent, and their terms can be rearranged arbitrarily. By the definition of the Riemann zeta function $\zeta(s)$, it follows that

$$\zeta'(s) = -\sum_{n\geq 1} \frac{\log n}{n^s} \quad \left(\zeta' \text{ denoting } \frac{d\zeta}{ds}\right).$$

We have

$$-\frac{\zeta'}{\zeta}(s) = \sum_{m\geq 1} \frac{\Lambda(m)}{m^s}.$$

[5] Throughout this subsection, r will always stand for a real number, such that $1 \leq r \leq \sqrt{N}$, unless otherwise stated.

We define

$$D_1(s) := \sum_{n \le r} \frac{\Lambda(n)}{n^s}, \quad D_2(s) := \sum_{n \le r} \frac{\mu(n)}{n^s}.$$

It follows that

$$0 = -\zeta'(s)D_2(s) - \zeta(s)D_1(s)D_2(s) - \zeta(s)D_2(s)\left(-\frac{\zeta'}{\zeta}(s) - D_1(s)\right).$$

$$= \sum_{n \ge 1} \frac{1}{n^s} \left(\sum_{\substack{m \le r \\ d \cdot m = n}} (\log d)\, \mu(m) \right) - \sum_{n \ge 1} \frac{1}{n^s} \sum_{\substack{d \cdot m = n \\ m \le r^2}} \sum_{\substack{q \mid d \\ q \le r}} \mu(q)\Lambda(m)$$

$$- \sum_{n \ge 1} \frac{1}{n^s} \sum_{\substack{m \cdot d = n \\ r < m}} \left(\sum_{\substack{t \mid d \\ t \le r}} \mu(t) \right) \Lambda(m).$$

Therefore, equating the coefficients of the Dirichlet series above we obtain:

$$\sum_{\substack{d \cdot m = n \\ m \le r}} \mu(m) \log d - \sum_{\substack{d \cdot m = n \\ m \le r^2}} \left(\sum_{\substack{q \mid d \\ q \le r}} \mu(q)\Lambda(m) \right) - \sum_{r < m} \left(\sum_{\substack{d \mid m \\ d \le r}} \mu(d) \right) \Lambda(m) = 0 \qquad (*)$$

From (*) by multiplying by $e(xn)$ and adding

$$\sum_{r < n \le N} \Lambda(n)e(xn)$$

we obtain

$$\sum_{r < n \le N} \Lambda(n)e(xn) = \sum_{1 \le n \le N} e(xn) \sum_{\substack{m \le r \\ d \cdot m = n}} (\log d)\mu(m)$$

$$- \sum_{1 \le n \le N} e(xn) \sum_{\substack{m \le r^2 \\ d \cdot m = n}} \left(\sum_{\substack{q \mid d \\ q \le r}} \mu(q) \right) \Lambda(m)$$

$$- \left(\sum_{1 \le n \le N} e(xn) \sum_{\substack{d \cdot m = n \\ r < m}} \left(\sum_{\substack{q \mid m \\ q \le r}} \mu(q) \right) \Lambda(m) - \sum_{r \le n \le N} \Lambda(n)e(xn) \right).$$

For $d \le r$, we have

$$\sum_{\substack{q|d \\ q \le r}} \mu(q) = \begin{cases} 1, & \text{if } d = 1 \\ 0, & \text{otherwise} \end{cases}$$

and therefore the proof of Lemma 3.6 is finished. □

Lemma 3.7 *Let*

$$A := \sum_{d \le r} \sum_{m \le \frac{N}{d}} \log m \cdot \mu(d) \cdot e(xdm)$$

Then

$$|A| \ll \log N \sum_{d \le r^2} \min \left\{ \frac{1}{[xd]}, \frac{N}{d} \right\},$$

where $[y] = \min_{k \in \mathbb{Z}} |y - k|$.

Proof We have

$$\sum_{1 \le m \le N/d} (\log m) e(xdm) = \sum_{1 \le m \le N/d} e(xdm) \int_1^m \frac{du}{u} = \int_1^{N/d} \left(\sum_{u < m \le N/d} e(xdm) \right) \frac{du}{u}.$$

Therefore,

$$\begin{aligned} |A| &\le \sum_{d \le r} \left| \sum_{m \le N/d} (\log m) e(xdm) \right| \\ &\le \sum_{d \le r} \min \left\{ \sum_{m \le N/d} \log m \,, \left| \int_1^{N/d} \sum_{u < m \le N/d} e(xdm) \frac{du}{u} \right| \right\} \\ &\ll \sum_{d \le r} \min \left\{ \frac{N}{d} \log N \,, \frac{\log N}{[xd]} \right\} \end{aligned}$$

Thus the proof of the lemma now follows. □

Lemma 3.8 *Let*

$$B := \sum_{d \le r^2} \sum_{m \le \frac{N}{d}} \sum_{\substack{q|d \\ q \le r \\ d \le rq}} \mu(q) \Lambda\left(\frac{d}{q}\right) e(xdm) .$$

Then

$$|B| \ll \log N \sum_{d \le r^2} \min\left\{\frac{1}{[xd]}, \frac{N}{d}\right\} .$$

Proof We have

$$|B| \le \sum_{d \le r^2} \left| \sum_{\substack{q|d \\ q \le r \\ d \le rq}} \Lambda\left(\frac{d}{q}\right) \right| \left| \sum_{m \le \frac{N}{d}} e(xdm) \right|$$

$$\le \sum_{d \le r^2} \sum_{q|d} \Lambda(q) \left| \sum_{m \le \frac{N}{d}} e(xdm) \right| .$$

By Theorem 1.13, we know that

$$\sum_{q|d} \Lambda(q) = \log d .$$

In addition, by Lemma 1.17 it follows that

$$\left| \sum_{m \le \frac{N}{d}} e(xdm) \right| \le \min\left\{\frac{1}{[xd]}, \frac{N}{d}\right\} .$$

Hence, we obtain

$$|B| \le \sum_{d \le r^2} \left((\log d) \cdot \min\left\{\frac{1}{[xd]}, \frac{N}{d}\right\} \right),$$

and therefore

$$|B| \ll \log N \sum_{d \le r^2} \min\left\{\frac{1}{[xd]}, \frac{N}{d}\right\} ,$$

which proves the lemma. □

Lemma 3.9 *Let*

$$C := \sum_{r<d\le N} \sum_{r<m\le \frac{N}{d}} \sum_{\substack{q|d \\ q\le r}} \mu(q)\Lambda(m)e(xdm)\,.$$

Then

$$|C| \ll \sum_{i=1}^{t} \left((2^i r \log^5 N) \sum_{r<d\le \frac{N}{2^i r}} \left(2^i r + \sum_{1\le s\le \frac{N}{2^i r}} \min\left\{ \frac{1}{[xs]}, \frac{N}{s} \right\} \right) \right)^{1/2},$$

where

$$t = \left\lfloor \frac{\log(N/r^2)}{\log 2} \right\rfloor .$$

Proof If we observe the indices under the first two sums of the definition of C, we see that

$$r < d \le N$$

and

$$r < m \le \frac{N}{d}\,. \tag{1}$$

But, for $N/r < d \le N$ it holds $N/d < r$ and thus (1) does not hold true. In that case, the second sum of C does not contain any terms. Therefore, it is evident that

$$C = \sum_{r<d\le \frac{N}{r}} \sum_{r<m\le \frac{N}{d}} \sum_{\substack{q|d \\ q\le r}} \mu(q)\Lambda(m)e(xdm)\,.$$

Let

$$D(r) = \sum_{r<m\le \frac{N}{d}} \sum_{\substack{q|d \\ q\le r}} \mu(q)\Lambda(m)e(xdm)\,.$$

Then, we can write

$$C = \sum_{r<d\le 2r} D(r) + \sum_{2r<d\le 4r} D(r) + \cdots + \sum_{2^t r<d\le 2^{t+1} r} D(r)\,, \tag{2}$$

where t is an integer, such that

$$2^t r < \frac{N}{r} \leq 2^{t+1} r .$$

From,

$$2^t r < \frac{N}{r} \leq 2^{t+1} r ,$$

we have

$$2^t < \frac{N}{tr^2} \leq 2^{t+1} .$$

Thus,

$$t < \frac{\log \left(N/r^2\right)}{\log 2} .$$

However, for $0 \leq i \leq t$, we have by the definition of $D(r)$ that

$$\left| \sum_{2^i r < d \leq 2^{i+1} r} D(r) \right|^2 = \left| \sum_{2^i r < d \leq 2^{i+1} r} \left(\sum_{r < m \leq \frac{N}{d}} \left(\sum_{\substack{q|d \\ q \leq r}} \mu(q) \right) \Lambda(m) e(xdm) \right) \right|^2$$

and by the Cauchy-Schwarz-Buniakowsky inequality we obtain

$$\left| \sum_{2^i r < d \leq 2^{i+1} r} D(r) \right|^2$$

$$\leq \left(\sum_{2^i r < d \leq 2^{i+1} r} \left| \sum_{\substack{q|d \\ q \leq r}} \mu(q) \right|^2 \right) \cdot \left(\sum_{2^i r < d \leq 2^{i+1} r} \left| \sum_{r < m \leq \frac{N}{d}} \Lambda(m) e(xdm) \right|^2 \right) \qquad (3)$$

Because of the fact that $|\mu(q)| \leq 1$, we have

$$\sum_{2^i r < d \leq 2^{i+1} r} \left| \sum_{\substack{q|d \\ q \leq r}} \mu(q) \right|^2 \leq \sum_{2^i r < d \leq 2^{i+1} r} \left| \sum_{\substack{q|d \\ q \leq r}} 1 \right|^2$$

$$= \sum_{2^i r < d \leq 2^{i+1} r} \tau^2(d) .$$

By Theorem 1.18 and due to the fact that $2^i r \leq N$, we obtain

$$\sum_{2^i r < d \leq 2^{i+1} r} \left| \sum_{\substack{q \mid d \\ q \leq r}} \mu(q) \right|^2 \ll (2^i r) \log^3(2^i r)$$

$$\leq (2^i r) \log^3 N .$$

Hence, by the above relation and (3), we get

$$\left| \sum_{2^i r < d \leq 2^{i+1} r} D(r) \right|^2 \ll (2^i r) \log^3 N \sum_{2^i r < d \leq 2^{i+1} r} \left| \sum_{r < m \leq \frac{N}{d}} \Lambda(m) e(xdm) \right|^2 \quad (4)$$

In addition, for the remaining sums in (4), we can write

$$\sum_{2^i r < d \leq 2^{i+1} r} \left| \sum_{r < m \leq \frac{N}{d}} \Lambda(m) e(xdm) \right|^2$$

$$= \sum_{2^i r < d \leq 2^{i+1} r} \left(\sum_{r < m \leq \frac{N}{d}} \Lambda(m) e(xdm) \cdot \sum_{r < s \leq \frac{N}{d}} \Lambda(s) e(-xds) \right)$$

$$\leq \sum_{r < m \leq \frac{N}{2^i r}} \left(\sum_{r < s \leq \frac{N}{2^i r}} \Lambda(m) \Lambda(s) \left| \sum_{2^i r < d \leq \min\{2^{i+1} r, \frac{N}{m}, \frac{N}{s}\}} e(xd(m-s)) \right| \right) .$$

However, $\Lambda(m)$, $\Lambda(s) \leq \log N$, for every $x < n$, $s \leq N/2^i r$. Thus,

$$\Lambda(m) \Lambda(s) \leq \log^2 N .$$

In addition, by Lemma 1.17, we have

$$\left| \sum_{2^i r < d \leq 2^{i+1} r} e(x(m-s)d) \right| \leq \min \left\{ \frac{1}{[x(m-s)]}, 2^i r \right\} .$$

Therefore, we obtain

$$\sum_{2^i r<d\le 2^{i+1}r}\left|\sum_{r<m\le \frac{N}{d}}\Lambda(m)e(xdm)\right|^2 \le \log^2 N\sum_{r<m\le \frac{N}{2^i r}}\sum_{r<s\le \frac{N}{2^i r}}\min\left\{\frac{1}{[x(m-s)]},2^i r\right\}$$
$$\ll \log^2 N\sum_{r<m\le \frac{N}{2^i r}}\sum_{0<s\le \frac{N}{2^i r}}\min\left\{\frac{1}{[xs]},2^i r\right\}$$
$$= \log^2 N\sum_{r<m\le \frac{N}{2^i r}}\left(2^i r+\sum_{1<s\le \frac{N}{2^i r}}\min\left\{\frac{1}{[xs]},2^i r\right\}\right).$$

Hence,

$$\sum_{2^i r<d\le 2^{i+1}r}\left|\sum_{r<m\le \frac{N}{d}}\Lambda(m)e(xdm)\right|^2 \le \log^2 N\sum_{r<m\le \frac{N}{2^i r}}\left(2^i r+\sum_{1<s\le \frac{N}{2^i r}}\min\left\{\frac{1}{[xs]},\frac{N}{s}\right\}\right).$$

By the above relation and (4), we obtain

$$\left|\sum_{2^i r<d\le 2^{i+1}r}D(r)\right| \ll \left((2^i r)\log^5 N\sum_{r<m\le \frac{N}{2^i r}}\left(2^i r+\sum_{1<s\le \frac{N}{2^i r}}\min\left\{\frac{1}{[xs]},\frac{N}{s}\right\}\right)\right)^{1/2}.$$

By the above relation and (2), it is evident that

$$|C| \ll \sum_{i=1}^{t}\left((2^i r\log^5 N)\sum_{r<d\le \frac{N}{2^i r}}\left(2^i r+\sum_{1\le s\le \frac{N}{2^i r}}\min\left\{\frac{1}{[xs]},\frac{N}{s}\right\}\right)\right)^{1/2}.$$

This completes the proof of Lemma 3.9. □

Corollary 3.10 *We have the following estimate*

$$|C| \ll \log^4 N\cdot\left(\frac{N}{\sqrt{r}}+\frac{N}{\sqrt{q}}+\sqrt{Nq}\right).$$

Proof By the previous lemma and Lemma 3.5, it follows that

$$
\begin{aligned}
|C| &\ll \sum_{1\le i\le t}\left(N\log^5 N \sum_{r<d\le \frac{N}{2^i r}}\left(2^i r+\left(\frac{N}{q}+\frac{N}{2^i r}+q\right)\log N\right)\right)^{1/2} \\
&= \sum_{1\le i\le t}\left(N\log^6 N\left(2^i r+\frac{N}{q}+\frac{N}{2^i r}+q\right)\right)^{1/2} \\
&\ll \sum_{1\le i\le t}\sqrt{N}\log^3 N\left(2^i r+\frac{N}{q}+\frac{N}{2^i r}+q\right)^{1/2} \\
&\le \sum_{1\le i\le t}\sqrt{N}\log^3 N\left(\sqrt{2^i r}+\sqrt{N/q}+\sqrt{N/2^i r}+\sqrt{q}\right).
\end{aligned}
$$

However, we have shown that

$$
t=\left\lfloor \frac{\log\left(N/r^2\right)}{\log 2}\right\rfloor .
$$

Thus,

$$
t \ll \log N ,
$$

which implies that

$$
\begin{aligned}
|C| &\ll \sqrt{N}\log^4 N\left(\sqrt{2^i r}+\sqrt{N/q}+\sqrt{N/2^i r}+\sqrt{q}\right) \\
&= \log^4 N\left(\sqrt{2^i rN}+\frac{N}{\sqrt{q}}+\frac{N}{\sqrt{2^i r}}+\sqrt{Nq}\right).
\end{aligned}
$$

But, since $r \le 2^i r \le N/r$, we obtain

$$
|C| \ll \log^4 N\cdot\left(\frac{N}{\sqrt{r}}+\frac{N}{\sqrt{q}}+\sqrt{Nq}\right),
$$

which proves the corollary. □

Lemma 3.11 *Let r be a real number, such that*

$$
\frac{N}{\sqrt{r}} = r^2 ,
$$

then

$$
\sum_{r<k\le N}\Lambda(k)e(xk) \ll \log^4 N\left(N^{4/5}+\frac{N}{\sqrt{q}}+\sqrt{Nq}\right).
$$

Proof For $N/\sqrt{r} = r^2$, by the previous corollary, we get

$$|C| \ll \log^4 N \left(r^2 + \frac{N}{\sqrt{q}} + \sqrt{Nq} \right) .$$

From Lemmas 3.5, 3.7 and 3.8, we also get:

$$|A| \, , \; |B| \ll \log^4 N \left(r^2 + \frac{N}{\sqrt{q}} + \sqrt{Nq} \right) .$$

Therefore, we have

$$|A| \, , \; |B| \, , \; |C| \ll \log^4 N \left(r^2 + \frac{N}{\sqrt{q}} + \sqrt{Nq} \right)$$

or

$$|A| \, , \; |B| \, , \; |C| \ll \log^4 N \left(N^{4/5} + \frac{N}{\sqrt{q}} + \sqrt{Nq} \right) . \tag{1}$$

But, by Lemma 3.6, we know that

$$\sum_{r<k\leq N} \Lambda(k)e(xk) = A - C - B . \tag{2}$$

Thus, by (1) and (2) we obtain that

$$\sum_{r<k\leq N} \Lambda(k)e(xk) \ll \log^4 N \left(N^{4/5} + \frac{N}{\sqrt{q}} + \sqrt{Nq} \right) .$$

This completes the proof of the lemma. □

Theorem 3.12 *(a) If $x \in \mathfrak{m}$, then*

$$f(x) \ll N \log^{5-D/2} N ,$$

where D is a positive constant, such that $D > 10$.[6]
(b) If c is a positive constant, such that $c \leq D/2 - 5$, then

$$\int_{\mathfrak{m}} f^3(x)e(-xN)dx \ll \frac{N^2}{\log^c N} .$$

[6]We mentioned the constant D in Theorem 1.24 (Siegel-Walfisz Theorem).

Proof (a) We have

$$\begin{aligned}\left|f(x)-\sum_{r<k\le N}\Lambda(k)e(xk)\right| &= \left|\sum_{p\le r}\log p\cdot e(xp)+\sum_{r<p\le N}\log p\cdot e(xp)-\sum_{r<k\le N}\Lambda(k)e(xk)\right|\\ &\le \left|\sum_{p\le r}\log p\cdot e(xp)\right|+\left|\sum_{r<p\le N}\log p\cdot e(xp)-\sum_{r<k\le N}\Lambda(k)e(xk)\right|\\ &= \sum_{p\le r}\log p+\left|\sum_{\substack{r<k\le N\\ k=p^m\\ m\ge 2}}\log p\cdot e(xp)\right| .\end{aligned}$$

Thus,

$$\left|f(x)-\sum_{r<k\le N}\Lambda(k)e(xk)\right|\le\sum_{p\le r}\log p+\sum_{\substack{r<k\le N\\ k=p^m\\ m\ge 2}}\log p . \tag{1}$$

But, as far as the second summand is concerned, we observe that

$$r<p^m\le N\,,\quad\text{or}\ p\le\sqrt[m]{N}\,,\quad m\ge 2\,.$$

For

$$m=\frac{\log N}{\log 2}$$

we get

$$p^m\ge 2^m=2^{\log N/\log 2}=e^{\log 2\frac{\log N}{\log 2}}=e^{\log N}=N$$

or

$$p^m\ge N\,.$$

Therefore, from (1) and the above observation, it follows that

$$\left|f(x)-\sum_{r<k\le N}\Lambda(k)e(xk)\right|\le\sum_{p\le r}\log p+\sum_{2\le m\le\left\lfloor\frac{\log N}{\log 2}\right\rfloor}\ \sum_{p^m\le N}\log p\,. \tag{2}$$

However,

$$\sum_{2\le m\le\left\lfloor\frac{\log N}{\log 2}\right\rfloor}\sum_{p^m\le N}\log p\le\sum_{p^2\le N}\log p+\log N\sum_{3\le m\le\left\lfloor\frac{\log N}{\log 2}\right\rfloor}\sum_{p^m\le N}1$$

$$\le\sum_{p^2\le N}\log p+\log N\sum_{3\le m\le\left\lfloor\frac{\log N}{\log 2}\right\rfloor}\sum_{p^3\le N}1$$

$$=\sum_{p^2\le N}\log p+\log N\left(\sum_{3\le m\le\left\lfloor\frac{\log N}{\log 2}\right\rfloor}1\right)\cdot\left(\sum_{p^3\le N}1\right).$$

But, since $\log 2>1/2$, we obtain

$$\sum_{2\le m\le\left\lfloor\frac{\log N}{\log 2}\right\rfloor}\sum_{p^m\le N}\log p\le\sum_{p^2\le N}\log p+\log N\left(\sum_{3\le m\le\left\lfloor\frac{\log N}{\log 2}\right\rfloor}1\right)\cdot\left(\sum_{p^3\le N}1\right)$$

$$\le\log N\left(\sum_{p^2\le N}1+2\log N\sum_{p^3\le N}1\right).$$

By the above relation and (2), we obtain

$$\left|f(x)-\sum_{r<k\le N}\Lambda(k)e(xk)\right|\le\log N\left(\sum_{p\le r}1+\sum_{p\le\sqrt{N}}1+2\log N\sum_{p\le\sqrt[3]{N}}1\right) \tag{3}$$

However, by Chebyshev's inequality, we know that for every positive integer n, where $n\ge 2$, it holds

$$\frac{1}{6}\cdot\frac{n}{\log n}<\pi(n)<6\cdot\frac{n}{\log n}.$$

Therefore, it is evident that

$$\sum_{p\le r}1\ll\frac{r}{\log r},\quad\sum_{p\le\sqrt{N}}1\ll\frac{\sqrt{N}}{\log\sqrt{N}},\quad\sum_{p\le\sqrt[3]{N}}1\ll\frac{\sqrt[3]{N}}{\log\sqrt[3]{N}}.$$

In addition, we have

$$\log N = 2 \cdot \frac{1}{2} \log N = 2 \log \sqrt{N} \ll \log \sqrt{N}$$

and similarly

$$\log N \ll \log \sqrt[3]{N} .$$

Hence, by the above arguments and (3), we get

$$\left| f(x) - \sum_{r<k\leq N} \Lambda(k)e(xk) \right| \ll r + \sqrt{N} \log N + 2\sqrt[3]{N} \log N$$

$$\ll \sqrt{N} \log N . \tag{4}$$

But, by Lemma 3.11, we know that

$$\sum_{r<k\leq N} \Lambda(k)e(xk) \ll \log^4 N \left(N^{4/5} + \frac{N}{\sqrt{q}} + \sqrt{Nq} \right) .$$

Thus, by (4), it follows that

$$f(x) \ll \log^5 N \left(N^{4/5} + \frac{N}{\sqrt{q}} + \sqrt{Nq} \right)$$

or

$$f(x) \ll N \log^5 N \left(N^{-1/5} + \frac{1}{\sqrt{q}} + \sqrt{\frac{q}{N}} \right) . \tag{5}$$

By Dirichlet's Approximation Theorem (see Theorem 1.22 for a proof), we know that for any real number x and natural number n, there exists an integer q, such that $0 < q \leq n$, and an integer a relatively prime to b, for which it holds

$$\left| x - \frac{a}{q} \right| < \frac{1}{nq} .$$

Since Dirichlet's Approximation Theorem holds for every $x \in \mathbb{R}$ and $n \in \mathbb{N}$, let us assume that $x \in \mathfrak{m}$ and n is a natural number, such that $n \geq N/L$ and $n - 1 < N/L$, where $L = \log^D N$. Then, we have

$$\left| x - \frac{a}{q} \right| \leq \frac{L}{Nq} \leq \frac{L}{N} .$$

Therefore, by the definition of the Major arcs, it follows that $x \in \mathfrak{M}$. But, it is impossible for x to belong in both the Major and the Minor arcs.

Thus, since

$$\mathfrak{M} = \bigcup_{\substack{1 \leq q \leq L \\ (a,q)=1}} \mathfrak{M}_{(a,q)} ,$$

it is evident that it must hold $q > L$. Hence, it is clear that

$$L < q < \frac{N}{L} ,$$

for $x \in \mathfrak{m}$.
Consequently, by (5), we obtain

$$\begin{aligned} f(x) &\ll N \log^5 N \left(N^{-1/5} + \frac{2}{\sqrt{L}} \right) \\ &= N \log^5 N \left(N^{-1/5} + 2 \log^{-D/2} N \right) \\ &\ll N \log^{5-D/2} N . \end{aligned}$$

(b) We proved above that

$$f(x) \ll N \log^{5-D/2} N .$$

Therefore, we have

$$\int_{\mathfrak{m}} f^3(x) e(-xN) dx \leq \int_0^1 \left| f^2(x) \right| \left| f(x) \right| dx \ll N \log^{5-D/2} N \int_0^1 \left| f^2(x) \right| dx \tag{1}$$

However,

$$\begin{aligned} \int_0^1 \left| f^2(x) \right| dx &= \int_0^1 f(x) f(-x) dx \\ &= \int_0^1 \sum_{p_1 \leq N} \log p_1 \cdot e(xp_1) \sum_{p_2 \leq N} \log p_2 \cdot e(-xp_2) dx \\ &= \sum_{p_1 \leq N} \log p_1 \sum_{p_2 \leq N} \log p_2 \int_0^1 e((p_1 - p_2)x) dx . \end{aligned}$$

But, by Lemma 1.2, we know that

$$\int_0^1 e((p_1 - p_2)x) dx = \begin{cases} 1, & \text{if } p_1 = p_2 \\ 0, & \text{if } p_1 \neq p_2 . \end{cases}$$

Thus, it is evident that

$$\int_0^1 |f^2(x)|\,dx \le \sum_{p\le N} \log^2 p \le \sum_{p\le N} \log^2 N$$
$$= \pi(N)\log^2 N\,.$$

Hence, by (1) and Chebyshev's inequality, we obtain

$$\int_{\mathfrak{m}} f^3(x)e(-xN)dx \ll N\log^{5-D/2} N \cdot \log^2 N \cdot \frac{N}{\log N}$$
$$= N^2 \log^{6-D/2} N$$
$$= \frac{N^2}{\log^c N}\,.$$

This proves Theorem 3.12. □

3.3 Putting It All Together

In this section we use the results obtained in the previous sections in order to prove Vinogradov's theorem.

Theorem 3.13 (VINOGRADOV'S THEOREM) *There exists a natural number N_0, such that every odd positive integer N with $N \ge N_0$, can be represented as the sum of three prime numbers.*

Proof Recall that by the arguments presented in the section related to the Circle Method, in order to prove Vinogradov's theorem it suffices to prove that

$$\overline{R}_N(m,3) = \sum_{\substack{m=p_1+p_2+p_3\\ p_i\le N}} \log p_1 \cdot \log p_2 \cdot \log p_3 \gg N^2\,.$$

However, we have

$$\overline{R}_N(m,3) = \int_{\mathfrak{M}} f^3(x)e(-mx)dx + \int_{\mathfrak{m}} f^3(x)e(-mx)dx\,.$$

In addition, by Theorems 3.4 and 3.12, we know that

$$\int_{\mathfrak{M}} f^3(x)e(-xN)dx - \frac{N^2}{2}\,G(N) \ll N^2 \log^{-D/2} N$$

and

$$\int_{\mathfrak{m}} f^3(x)e(-xN)dx \ll N^2 \log^{-c} N\,,$$

where c is a positive constant, such that $c \leq D/2 - 5$ and

$$G(N) = \sum_{q=1}^{+\infty} \frac{\mu(q)c_q(N)}{\phi(q)^3}\,.$$

Therefore,

$$\overline{R}_N(m,3) - \frac{N^2}{2}\,G(N) \ll \frac{N^2}{\log^w N}\,, \tag{1}$$

where w is a positive constant, such that $w \leq D/2 - 5$.
Generally, for any Dirichlet series with coefficients $f(n)$, where $f(n)$ is a multiplicative arithmetic function, by Theorem 1.5, it holds

$$\sum_{n=1}^{+\infty} \frac{f(n)}{n^s} = \prod_p \left(\sum_{n=0}^{+\infty} \frac{f(p^n)}{p^{ns}} \right),$$

where the product extends over all prime numbers p.
Therefore, for $s = 0$, we get

$$\sum_{n=1}^{+\infty} f(n) = \prod_p \left(\sum_{n=0}^{+\infty} f(p^n) \right).$$

In our case, since the arithmetic function

$$\frac{\mu(q)c_q(n)}{\phi(q)^3}$$

is multiplicative, we can write

$$G(N) = \sum_{q=1}^{+\infty} \frac{\mu(q)c_q(n)}{\phi(q)^3} = \prod_p \left(\sum_{n=0}^{+\infty} \frac{\mu(p^n)c_{p^n}(N)}{\phi(p^n)^3} \right).$$

However, for $n > 1$ we have $\mu(p^n) = 0$. Thus,

$$G(N) = \prod_p \left(\frac{\mu(1)c_1(N)}{\phi(1)^3} + \frac{\mu(p)c_p(N)}{\phi(p)^3} \right) = \prod_p \left(1 + \frac{(-1)c_p(N)}{(p-1)^3} \right) \tag{2}$$

But, by Lemma 1.16, we know that

$$c_p(N) = \sum_{d|(p,N)} \mu\left(\frac{p}{d}\right) d\,.$$

Hence, if $p \mid N$, then the only possible values of d are 1 and p. Thus,

$$c_p(N) = p - 1\,.$$

Similarly, if $p \nmid N$, we get

$$c_p(N) = -1\,.$$

Therefore, by (2), it follows that

$$\begin{aligned} G(N) &= \prod_{p|N}\left(1 - \frac{p-1}{(p-1)^3}\right)\prod_{p\nmid N}\left(1 + \frac{1}{(p-1)^3}\right) \\ &= \prod_{p|N}\left(1 - \frac{1}{(p-1)^2}\right)\prod_{p\nmid N}\left(1 + \frac{1}{(p-1)^3}\right). \end{aligned}$$

For odd integer N, we have

$$1 - \frac{1}{(p-1)^2} > 0$$

for all $p \mid N$.
Furthermore, the infinite series

$$\sum_p \frac{1}{(p-1)^2} \quad \text{and} \quad \sum_p \frac{1}{(p-1)^3}$$

are absolutely convergent, and thus, the infinite products

$$\prod_{p|N}\left(1 - \frac{1}{(p-1)^2}\right) \quad \text{and} \quad \prod_{p\nmid N}\left(1 + \frac{1}{(p-1)^3}\right)$$

are bounded from below and from above by bounds that are independent from N. Hence, Vinogradov's theorem is now proved. □

The Ternary Goldbach Problem with a Prime and Two Isolated Primes

1 Introduction

In this chapter, we present the result of Maier and Rassias [37] that under the assumption of the Generalized Riemann Hypothesis each sufficiently large odd integer can be expressed as the sum of a prime and two isolated primes.

The work of Vinogradov [66] has been extended by many authors. Recently, Helfgott [24] proved the result that all odd integers greater than five have this property. Wirsing [69], motivated by earlier work of Erdős and Nathanson [13] on sums of squares, considered the question of whether one could find thin subsets S of primes, which are still sufficient to obtain all large odd integers as sums of three of them and obtained the answer that there exist such sets S with

$$\sum_{p \leq x, p \in S} 1 \ll (x \log x)^{1/3} .$$

Wirsing's result used probabilistic methods and did not lead to a subset of the primes, which is constructive. Wolke in an Oberwolfach conference in 1986 suggested to find more familiar thin sets of primes. This was achieved by Balog and Friedlander [2] who merged the result of Vinogradov with that of Piatetski-Shapiro [42], who considered sets

$$P_C = \{p \mid p \text{ prime}, \ p = [n^C]\}$$

with a fixed constant $C > 1$.

Piatetski-Shapiro [42] proved that

$$\sum_{\substack{n \leq x \\ [n^C] prime}} 1 = (1 + o(1)) \frac{x}{C \log x}$$

M.T. Rassias, *Goldbach's Problem*, DOI 10.1007/978-3-319-57914-6_3

holds in the range $1 < C < 12/11$, which later was improved by many authors.
Balog and Friedlander proved ([2], Corollary 1):

For any fixed C with $1 < C < 21/20$ every sufficiently large odd integer can be written as the sum of three primes from P_C.

They also prove more general results in which the three primes are taken from sets P_C with possibly different values of C.
Here, we consider another special set of primes: **isolated primes**, which we define as follows:

Definition 3.1.1 Let $g : \mathbb{N} \to [1, \infty)$ be a monotonically increasing function with $g(n) \to \infty$ for $n \to \infty$. We say that a prime p is g**-isolated**, if m is composite for all positive integers m with

$$0 < |p - m| \leq (\log p)g(p) .$$

This concept of isolation is closely linked to the question of large gaps between primes. Let p_n be the nth prime. For which functions g do we have:

$$p_{n+1} - p_n \geq (\log p_n)g(p_n) \tag{1.1}$$

infinitely often?
After Westzynthius [68] had shown that (1.1) holds for g being an arbitrarily large positive constant, further progress was achieved by variations of the Erdős-Rankin method [12, 53]. Rankin [53] could show that (1.1) holds with

$$g(p_n) = C\,\frac{\log_2 p_n \log_4 p_n}{\log_3^2 p_n}\,, \tag{1.2}$$

(C a positive constant, $\log_1 n = \log n$, $\log_k n = \log(\log_{k-1} n)$).
For a long period of time, the improvements of the result (1.2) only involved the constant C [36, 44, 54, 59]. A famous prize problem of Paul Erdős was the improvement of the order of magnitude of the function g. This was achieved only recently [15, 16, 38]. The latest result [16] is:

$$g(p_n) = C\,\frac{\log_2 p_n \log_4 p_n}{\log_3 p_n}\,.$$

We shall use this function g in our definition of isolated primes, which in the sequel we shall simply call isolated.

Definition 3.1.2 Let C be a fixed positive constant, which we assume to be sufficiently small. We say a prime number p is **isolated** if $\log_4 p \geq 1$ and m is composite for all integers m with

$$0 < |p - m| \leq C\,(\log p)\,\frac{\log_2 p\,\log_4 p}{\log_3 p}\,.$$

The result of Maier and Rassias [37] is the following:

Theorem 3.1.3 *Assume the Generalized Riemann Hypothesis. Then each sufficiently large odd integer is a sum of a prime and two isolated primes.*

Remark. There are several challenges. It is very likely true that each sufficiently large odd integer is the sum of three isolated primes. It would be worthwhile to find a proof of this fact, possibly with a function g of smaller order of magnitude. It also remains the problem to remove the assumption of the Generalized Riemann Hypothesis.

2 Construction of the Isolated Residue Class

Let p_1, p_2 be two isolated primes in the representation $N = p_1 + p_2 + p_3$. In this section, we shall present the construction of a modulus P^* being a product of many small prime numbers. We also describe the construction of an "isolated" residue-class $u_0 (\text{mod } P^*)$, such that

$$(u_0, P^*) = (N - 2u_0, P^*) = 1 \text{ and } (m, P^*) > 1$$

for $m \neq u_0$ if $|u_0 - m|$ is small.
The proof of Theorem 3.1.3 will then be concluded in the next section by the circle method, choosing p_1 and p_2 from the residue-class $u_0 (\text{mod } P^*)$ and p_3 from the residue-class $(N - 2u_0) \bmod P^*$. From

$$(p_1 + n, P^*) = (p_2 + n, P^*) = (u_0 + n, P^*) > 1\,,$$

it then follows that p_1 and p_2 are isolated in the sense as defined in the previous section. More specifically, the proof of the following theorem is given in this section:

Theorem 3.2.1 *Let $C > 0$ and $D > 0$ be fixed, C sufficiently small and D sufficiently large. Let $N \geq N(C, D)$ be a sufficiently large odd integer. Then there is a positive integer P^* with*

$$(P^*)^{100} \leq N \leq (P^*)^D$$

and an integer u_0 with

$$(P^*)^C \leq u_0 \leq P^* ,$$

such that

$$(u_0, P^*) = (N - 2u_0, P^*) = 1 \ \ and \ \ (m, P^*) > 1$$

for

$$0 < |u_0 - m| \leq C(\log N) \ \frac{\log_2 N \ \log_4 N}{\log_3 N} .$$

The construction is mainly based on the ideas of the paper [16]. Also, the definitions are mainly taken from [16].
Let c_1, c_2 be fixed positive constants to be specified later. Also, the constants $c_3, c_4, \ldots$ will be positive constants. They will depend only on c_1, c_2. Set

$$x = c_1 \ \log N \ , \ y = c_2 x \ \frac{\log x \ \log_3 x}{\log_2 x} \ , \ z = x^{\log_3 x/(4\log_2 x)} \qquad (2.1)$$

Definition 3.2.2 We introduce the three disjoint sets of primes

$$S = \{s \text{ prime} \mid \log^{20} x < s \leq z\}, \mathcal{P} = \{p \text{ prime} \mid x/2 < p \leq x\} \qquad (2.2)$$
$$Q = \{q \text{ prime} \mid x < q \leq y\}$$

For vectors $\vec{a} = (a_s \bmod s)_{s \in S}$, $\vec{b} = (b_p \bmod p)_{p \in \mathcal{P}}$, we define the sifted sets

$$S(\vec{a}) = \{n \in \mathbb{Z} \ : \ n \not\equiv a_s (\text{mod } s) \text{ for all } s \in S\}$$

and likewise

$$S(\vec{b}) = \{n \in \mathbb{Z} \ : \ n \not\equiv b_p (\text{mod } p) \text{ for all } p \in \mathcal{P}\}$$

Definition 3.2.3 For each prime $p \leq x$, we define:

$$d_p = \begin{cases} a_s \, , & \text{if } p = s \in S \\ b_p \, , & \text{if } p \in \mathcal{P} \\ 0 \, , & \text{for all other } p \end{cases}$$

Lemma 3.2.4 *Let $n \in (x, y]$ satisfy $n \not\equiv d_p (\text{mod } p)$ for all $p \leq x$. Then $n \in Q \cap S(\vec{a}) \cap S(\vec{b})$ or $n \in R$, where*

$$R = \{n \in (x, y] \; : \; p \mid n \Rightarrow p \leq z\}$$

is the set of z-smooth integers in $(x, y]$.

Proof Assume $n \not\equiv d_p (\text{mod } p)$ for all $p \leq x$ and $n \notin Q$. Assume there are two primes t_1, t_2 with $t_1 < t_2, t_2 > z$ and $t_1 \mid n, t_2 \mid n$.
Then, by (2.2) it follows:

$$n \geq t_1 t_2 \geq \frac{x}{2} (\log x)^{20} > y \, ,$$

a contradiction. Thus, since $n \notin Q$, we have $n \in R$. □

Lemma 3.2.5 *It holds*

$$\#R = O\left(\frac{x}{(\log x)^2}\right) .$$

Proof To estimate $\#R$, let

$$u = \frac{\log y}{\log z} .$$

Thus, from (2.1) one has

$$u = 4 \frac{\log_2 x}{\log_3 x} (1 + o(1)).$$

By standard estimates for smooth numbers (e.g. de Bruijn's theorem [5]) and (2.1), it follows:

$$\#R \ll y e^{-u \log u + O(u \log \log(u+2))} = O\left(\frac{y}{\log^3 x}\right) = O\left(\frac{x}{\log^2 x}\right) .$$

□

Lemma 3.2.6 *Let N be sufficiently large and suppose that x, y, z are given by (2.1). Then there are vectors $\vec{a} = (a_s \mod s)_{s \in S}$ and $\vec{b} = (b_p \mod p)_{p \in \mathcal{P}}$, such that*

$$\#(Q \cap S(\vec{a}) \cap S(\vec{b})) \ll \frac{x}{\log x} . \tag{2.3}$$

Proof This follows immediately from Theorem 2 of [16]. □

We now fix $\vec{a}$ and $\vec{b}$ satisfying (2.3). The following results from standard sieves (Brun's or Selberg's sieve) are needed. The following notations and results are borrowed from [18]. An exception is the set of primes, denoted by $\mathcal{P}$ in [18], which we denote by $\tilde{\mathcal{P}}$.

Definition 3.2.7 Let $\mathcal{A}$ be a finite set of integers, and let $\tilde{\mathcal{P}}$ be a set of primes. For a positive squarefree integer d composed only of primes of $\tilde{\mathcal{P}}$, let

$$\mathcal{A}_d = \{n \in \mathcal{A} \mid n \equiv 0 (\text{mod } d)\} .$$

Let z be a positive real number, and let $P(z)$ be a product of the primes in $\tilde{\mathcal{P}}$ that are smaller than z. Then, set

$$S(\mathcal{A}; \tilde{\mathcal{P}}, z) = |\{a \mid a \in \mathcal{A},\ (a, P(z)) = 1\}| . \tag{2.4}$$

Let ω be a multiplicative function defined for squarefree numbers with $\omega(p) = 0$ for $p \notin \tilde{\mathcal{P}}$. With $X > 1$ set:

$$R_d = |\mathcal{A}_d| - \frac{\omega(d)}{d} X . \tag{2.5}$$

Define also

$$W(z) = \prod_{p<z} \left(1 - \frac{\omega(p)}{p}\right) . \tag{2.6}$$

We introduce the conditions:

$$0 \le \frac{\omega(p)}{p} \le 1 - \frac{1}{A_1} \tag{Ω_1}$$

for some constant $A_1 \ge 1$ (i.e. independent of X).

$$\sum_{w \le p < z} \frac{\omega(p) \log p}{p} \le \kappa \log \frac{z}{w} + A_2 , \tag{$\Omega_2(\kappa)$}$$

if $2 \le w \le z$, where $\kappa > 0$ and $A_2 \ge 1$ are independent of z and w.

$$|R_d| \le L \left(\frac{X \log X}{d} + 1\right) A_0^{\prime\nu(d)} , \quad \text{for } \mu(d) \ne 0, (d, \overline{\mathcal{P}}) = 1, \tag{R_0}$$

where $L \ge 1$ and $A_0' \ge 1$ are independent of X,

$$\overline{\mathcal{P}} = \{p \in P \mid p \notin \mathcal{P}\} .$$

Let α be a constant with $0 < \alpha \le 1$.

Assume that for any $U \geq 1$, there exists a $C_0 > 0$, such that

$$\sum_{\substack{d<X^{\alpha}\log^{-C_0} X \\ (d,\overline{\mathcal{P}})=1}} \mu^2(d)|R_d| = O_U\left(\frac{X}{\log^{\kappa+U} X}\right) . \qquad (R_1(\kappa,\alpha))$$

Lemma 3.2.8 *Let $\mathcal{A}$ satisfy the conditions (Ω_1), $(\Omega_2(\kappa))$, (R_0), $(R_1(\kappa,\alpha))$. Let $X \geq z$ and write*

$$u = \frac{\log X}{\log z}.$$

Then

$$\begin{aligned} S(\mathcal{A};\tilde{\mathcal{P}},z) = &XW(z)\{1 + O(\exp\{-\alpha u(\log u - \log\log 3u - \log(x/\alpha) - 2)\}) \\ &+ O_U(L\log^{-U} X)\}, \end{aligned} \qquad (2.7)$$

where the O-constants may depend on U as well as on the constants A_1', A_1, A_2, κ and α.

Proof This is Theorem 2.5′ of [18]. □

Definition 3.2.9 Let

$$Q^* = \left\{ q \in Q \,\middle|\, \frac{y}{3} < q \leq \frac{2y}{3},\ p \mid N - 2q \Rightarrow p > \log^{20} x \right\} .$$

Lemma 3.2.10

$$\#Q^* \geq c_3 \frac{y}{\log x \, \log_2 x} .$$

Proof We apply Lemma 3.2.8 with

$$\mathcal{A} = \left\{ N - 2q \,:\, \frac{y}{3} < q \leq \frac{2y}{3} \right\},\ z = (\log x)^{20},$$

$$X = li\,\frac{2y}{3} - li\,\frac{y}{3},\ \omega(p) = \begin{cases} \frac{p}{p-1}, & \text{for } p > 2,\ p \nmid N \\ 0, & \text{for } p = 2,\ p \mid N \end{cases}$$

Then, the conditions (Ω_1), $(\Omega_2(\kappa))$ are satisfied with $A_1 = 2$, $\kappa = 1$. The existence of A_2 follows from Mertens' result

$$\sum_{p \leq x} \frac{\log p}{p} = \log x + O(1) .$$

By the Generalized Riemann Hypothesis, one has

$$|R_d| \leq X^{1/2} \log^3 X \,, \quad \text{for } d \leq X^{1/2}.$$

Then,

$$W(z) = \prod_{p<(\log x)^{20}} \left(1 - \frac{\omega(p)}{p}\right) = O\left(\frac{1}{\log\log x}\right).$$

Lemma 3.2.10 now follows from (2.7). □

Definition 3.2.11 For $w \geq 1$, $r, b \in \mathbb{N}$, let

$$\pi(w, r, b) = \#\{p \leq w \; : \; p \equiv b (\bmod r)\}.$$

Lemma 3.2.12 *For* $1 \leq r \leq b$, $(b, r) = 1$, *we have:*

$$\pi(w, r, b) = O\left(\frac{w}{\phi(r) \log(w/r)}\right).$$

Proof This is the Brun–Titchmarsh inequality (see [18], Theorem 3.8). □

Definition 3.2.13 Let $q \in Q^*$. Then, define

$$\mathcal{N}_1(q) = \{s \in S \mid q \equiv a_s (\bmod s)\}$$

$$\mathcal{N}_2(q) = \{p \in \mathcal{P} \mid q \equiv b_p (\bmod p)\}$$

$$\mathcal{N}_3(q) = \{p \leq x \mid N - 2q \equiv 0 (\bmod p)\}.$$

Lemma 3.2.14 *Let* $x \to \infty$. *Then, we have for all but* $o(|Q^*|)$ *primes* $q \in Q^*$:

$$\#\mathcal{N}_1(q) = O(\log x \; \log_2 x \; \log_3 x) \tag{2.8}$$

$$\#\mathcal{N}_2(q) = O((\log_2 x)^2) \tag{2.9}$$

$$\#\mathcal{N}_3(q) = O((\log_2 x)^3) \tag{2.10}$$

Proof From Lemma 3.2.12:

$$\sum_{q \in Q^*} \#\mathcal{N}_1(q) = \sum_{s \in S} \sum_{\substack{q \equiv a_s (\bmod s) \\ q \in Q^*}} 1 = O\left(\frac{y}{\log x} \sum_{s \in S} \frac{1}{s}\right) = O\left(\frac{y}{\log x} \frac{\log x \; \log_3 x}{\log_2 x}\right).$$

The relation (2.8) follows from Lemma 3.2.10.

We have

$$\sum_{q\in Q^*} \#\mathcal{N}_2(q) = \sum_{p\in\mathcal{P}} \sum_{\substack{q\equiv b_p(\bmod\ p)\\ q\in Q^*}} 1 = O\left(y\sum_{p\in\mathcal{P}}\frac{1}{p}\right) = o\left(\frac{y}{\log x}\right).$$

The relation (2.9) now follows from Lemma 3.2.10. It holds

$$\sum_{q\in Q^*} \#\mathcal{N}_3(q) = \sum_{(\log x)^{20}<p\leq x} \sum_{\substack{q\in Q^*\\ N-2q\equiv 0(\bmod\ p)}} 1 = O\left(y \sum_{(\log x)^{20}<p\leq x}\frac{1}{p}\right) = O(y\log\log x).$$

The relation (2.10) now follows from Lemma 3.2.10. □

Lemma 3.2.15 *There is a prime* $q_0 \in \left[\frac{y}{3}, \frac{2y}{3}\right]$ *and a subset*

$$\mathcal{P}^* \subset \{p \leq x \mid p \textit{ prime}\},$$

such that $q_0 \not\equiv d_p(\bmod\ p)$ *for* $p \in \mathcal{P}^*$, $p \nmid N - 2q_0$ *for* $p \in \mathcal{P}^*$.

$$\#\{n \in (x, y] \mid n \not\equiv d_p(\bmod\ p) \ \textit{for}\ p \in \mathcal{P}^*\} = O\left(\frac{x}{\log x}\right).$$

Proof Since the sets $\{n \equiv d_p(\bmod\ p)\}$ are arithmetic progressions with difference p, one has by Lemma 3.2.14:

$$\begin{aligned}\#\{n \in (x, y] \mid n \equiv a_s(\bmod\ s) \ \text{for some}\ s \in \mathcal{N}_1(q_0)\} &= O\left(\sum_{s\in\mathcal{N}_1(q_0)}\frac{y}{s}\right)\\ &= O\left(y(\log x)^{-20}\#\mathcal{N}_1(q_0)\right)\\ &= O(x(\log x)^{-17}),\end{aligned} \tag{2.11}$$

$$\begin{aligned}\#\{n \in (x, y] \mid n \equiv b_p(\bmod\ p) \ \text{for some}\ s \in \mathcal{N}_2(q_0)\} &= O(x\log x(\log_2 x)^2 z^{-1})\\ &= O(x(\log x)^{-19})\end{aligned} \tag{2.12}$$

and

$$\#\{n \in (x, y] \mid n \equiv d_p(\bmod\ p) \ \text{for some}\ p \in \mathcal{N}_3(q_0)\} = O(x(\log x)^{-19}). \tag{2.13}$$

We now define

$$\mathcal{P}^* = \{p \text{ prime}, \ p \le z\} \setminus \{\mathcal{N}_1(q_0) \cup \mathcal{N}_2(q_0) \cup \mathcal{N}_3(q_0)\}\,. \tag{2.14}$$

Let

$$N_1 = \#\{n \in (x, y] \mid n \not\equiv d_p (\text{mod } p) \text{ for } p \in \mathcal{P}^*\}$$

$$N_2 = \#\{n \in (x, y] \mid n \not\equiv d_p (\text{mod } p) \text{ for } p \in \mathcal{P}\}\,.$$

The difference $N_1 - N_2$ may be estimated by the number of integers counted in (2.11), (2.12) and (2.13). Thus,

$$N_1 - N_2 = O\left(x(\log x)^{-17}\right)\,.$$

By Lemmas 3.2.4, 3.2.5, 3.2.6, one has

$$N_1 = O\left(\frac{x}{\log x}\right)\,,$$

which concludes the proof of Lemma 3.2.15. □

The proof of Theorem 3.2.1 is now concluded.
Let $n_1, \ldots, n_k \in (x, y] \setminus \{q_0\}$ with $n_j \not\equiv d_p(\text{mod } p)$ for all $p \in \mathcal{P}^*$. By the prime number theorem, there is a $C_0 > 1$, such that

$$\pi(C_0 x) - \pi(x) \ge 2k\,.$$

Since $p \mid (n_j - q_0)$ for at most one prime $p \in (x, C_0 x]$, we may choose primes $\tilde{p}_1, \ldots, \tilde{p}_k \in (x, C_0 x]$, such that $n_j \not\equiv q_0(\text{mod } \tilde{p}_j)$ for $1 \le j \le k$.

Definition 3.2.16 We define

$$d_{\tilde{p}_j} = n_j, \quad \text{for } 1 \le j \le k.$$

Now set

$$P^* = \prod_{p \in \mathcal{P}^*} p \prod_{j=1}^{k} \tilde{p}_j \tag{2.15}$$

and determine v_0 by the conditions

$$\begin{aligned} &1 \le v_0 < P^* \\ &v_0 \equiv -d_p(\text{mod } p) \ \text{ for all } p \mid P^*\,. \end{aligned} \tag{2.16}$$

By the Chinese Remainder Theorem, v_0 is uniquely determined. We now define the isolated residue class (mod P^*) by

$$u_0 = v_0 + q_0 \,. \tag{2.17}$$

From (2.16), one sees that $p \mid v_0 + d_p$ for all $p \in \mathcal{P}^* \setminus \{q_0\}$. Thus, $(n, P^*) > 1$ for all $n \in (x, y] \setminus \{u_0\}$. Therefore, $|u_0 - m|$ is composite for all m with $0 < |m| \leq y/3$. By Definition 3.2.13 for $\mathcal{N}_3(q)$ and (2.13), one has:

$$(u_0, P^*) = (N - 2u_0, P^*) = 1 \,.$$

It follows

$$P^* \leq P(C_0 x) = \exp(C_0 x(1 + o(1))) \tag{2.18}$$

by the prime number theorem.
By the Definition 3.2.3, one has $d_p = 0$ for $z < p \leq x/2$. Therefore,

$$u_0 \geq \prod_{\substack{z<p\leq x/2 \\ p\notin \mathcal{N}_1(q_0)\cup\mathcal{N}_2(q_0)\cup\mathcal{N}_3(q_0)}} p \geq (P^*)^c$$

if c is chosen small enough.
The bound $N \geq (P^*)^{100}$ follows, if c_1 in (2.1) is chosen sufficiently small. The upper bound $N \leq (P^*)^D$ follows from Lemma 3.2.14. This concludes the proof of Theorem 3.2.1.

3 The Circle Method

Let N and P^* satisfy the conditions of Theorem 3.2.1, and let u_0 be the isolated residue-class mod P^*. It remains to be shown that there are primes p_1, p_2, p_3 with $p_1 \equiv p_2 \equiv u_0 (\text{mod } P^*)$, $p_3 \equiv N - 2u_0 (\text{mod } P^*)$ with

$$p_1 + p_2 + p_3 = N \,. \tag{3.1}$$

This will be achieved by the circle method.
In the sequel, the paper [34] is followed closely. The results and definitions are borrowed from there with slight modifications. They are formulated for general arithmetic progressions.

Definition 3.3.1 Let $R \leq N^{1/20}$, $P = R^3 L^{3C}$, $Q = NR^{-3}L^{-4C}$, $L = \log N$, with the constant C to be specified later. The Major arc of the circle method is defined as

$$E_1(R) = \bigcup_{q \le P} \bigcup_{\substack{a=1 \\ (a,q)=1}}^{q} \left[\frac{a}{q} - \frac{1}{qQ}, \frac{a}{q} + \frac{1}{qQ}\right] \tag{3.2}$$

$$E_2(R) = \left[\frac{1}{Q}, 1 + \frac{1}{Q}\right] - E_1(R) . \tag{3.3}$$

Since $2P < Q$, no two Major arcs intersect. Write $\alpha \in [0, 1]$ in the form

$$\alpha = \frac{a}{q} + \lambda, \ 1 \le a \le q, \ (a, q) = 1 . \tag{3.4}$$

It follows from Dirichlet's lemma on rational approximation that

$$E_2(R) = \left\{\alpha \ : \ P < q < Q, \ |\lambda| \le \frac{1}{qQ}\right\} .$$

Let $\Lambda(n)$ be the von Mangoldt function, $e(\alpha) = e^{2\pi i \alpha}$ and

$$S(\alpha, r, b) = \sum_{\substack{n \le N \\ n \equiv b (\bmod r)}} \Lambda(n) e(n\alpha) . \tag{3.5}$$

By orthogonality, one obtains

$$\sum_{\substack{(n_1,n_2,n_3)\in\mathbb{N}^3, \\ n_i \equiv b_i (\bmod r) \\ n_1+n_2+n_3=N}} \Lambda(n_1)\Lambda(n_2)\Lambda(n_3) = \int_0^1 S(\alpha, r, b_1)S(\alpha, r, b_2)S(\alpha, r, b_3)e(-N\alpha)d\alpha . \tag{3.6}$$

Lemma 3.3.2 *Let $A > 0$ be arbitrary and $\alpha \in E_2(R)$. If C is sufficiently large, then*

$$S(\alpha, r, b) \ll \frac{N}{r \log^A N} \tag{3.7}$$

uniformly for $\frac{1}{2}R \le r \le 2R$.

Proof By the explanation in Definition 3.3.1, we have

$$\alpha = \frac{a}{q} + \lambda \text{ with } P < q < Q, \ |\lambda| \le \frac{1}{qQ} .$$

By a result of Balog and Perelli [3], we have for $M \le N$ and with $h = (r, q)$:

$$\sum_{\substack{n \le M \\ n \equiv b(\bmod r)}} \Lambda(n) e\left(\frac{a}{q} n\right) \ll L^3 \left(\frac{hN}{rq^{1/2}} + \frac{q^{1/2} N^{1/2}}{h^{1/2}} + \frac{N^{4/5}}{r^{2/5}}\right).$$

This, together with the bound for R, P and Q given in Definition 3.3.1, implies that for arbitrary large constants $B > 0$, there is a constant C, such that

$$\sum_{\substack{n \le M \\ n \equiv b(\bmod r)}} \Lambda(n) e\left(\frac{a}{q} n\right) \ll \frac{N}{r \log^B N}, \quad \text{for } M \le N.$$

The result of Lemma 3.3.2 follows by partial summation, observing that $|\lambda| \le 1/qQ$. □

Definition 3.3.3 (*Definitions from* [34])
Let d, f, g, k, m be fixed positive integers and χ_g a Dirichlet character mod g. Let

$$G(d, f, m, \chi_g, k) = \sum_{\substack{n=1 \\ (n,k)=1 \\ n \equiv f(\bmod d)}}^{k} \chi_g(n) e\left(\frac{mn}{k}\right) \tag{3.8}$$

Remark. This is a generalization of the Gaussian sum

$$G(m, \chi) = \sum_{n=1}^{k} \chi(n) e\left(\frac{mn}{k}\right).$$

In the special case $g = k$, let

$$G(d, f, m, \chi) = G(d, f, m, \chi_k, k). \tag{3.9}$$

For positive integers r and q, let

$$h = (r, q). \tag{3.10}$$

Then, r, q and h can be written as

$$\begin{aligned} r &= \tilde{p}_1^{\alpha_1} \cdots \tilde{p}_s^{\alpha_s} r_0 \ \ (\tilde{p}_j, r_0) = 1 \\ q &= \tilde{p}_1^{\beta_1} \cdots \tilde{p}_s^{\beta_s} q_0 \ \ (\tilde{p}_j, q_0) = 1 \\ h &= \tilde{p}_1^{\gamma_1} \cdots \tilde{p}_s^{\gamma_s}, \ \text{ with prime numbers } \tilde{p}_i, \end{aligned} \tag{3.11}$$

where α_j, β_j and γ_j are positive integers with $\gamma_j = \min(\alpha_j, \beta_j)$, $j = 1, \ldots, s$. Define

$$h_1 = \tilde{p}_1^{\delta_1} \cdots \tilde{p}_s^{\delta_s}, \tag{3.12}$$

where $\delta_j = \alpha_j$ or 0 according as $\alpha_j = \gamma_j$ or not. Then, $h_1 \mid h$. Write

$$h_2 = \frac{h}{h_1}. \tag{3.13}$$

Then

$$h_1 h_2 = h, \ (h_1, h_2) = 1, \ \left(\frac{r}{h_1}, \frac{q}{h_2}\right) = 1. \tag{3.14}$$

Lemma 3.3.4 *Let $d \mid k$, $g \mid k$ and $(m, k) = (f, k) = 1$. Let also $\chi \bmod g$ be induced by the primitive character $\chi^* \bmod g^*$. Then*

$$|G(d, f, m, \chi_g, k)| \le g^{*1/2}.$$

Proof This is Lemma 3 of [34]. □

Lemma 3.3.5 *Let $d \mid k$ and $(m, k) = (f, k) = 1$. Let also $\chi^0 (\bmod k)$ be the principal character. Then for $(d, k/d) > 1$,*

$$G(d, f, m, \chi^0) = 0$$

and for $(d, k/d) = 1$,

$$G(d, f, m, \chi^0) = \mu\left(\frac{k}{d}\right) e\left(\frac{fmt}{d}\right),$$

where t is defined by $tk/d \equiv 1 (\bmod d)$.

Proof This is Lemma 7 of [34]. □

The basic identity for the asymptotic evaluation of the Major arcs is the following.

Lemma 3.3.6 *Let a, q, r be positive integers, and h, h_1, h_2 defined as in (3.10)–(3.13), such that (3.14) holds. Then*

$$S\left(\frac{a}{q} + \lambda, r, b\right) = \frac{1}{\phi(r/h_1)\phi(q/h_2)} \sum_{\xi (\bmod r/h_1)} \overline{\xi}(b) \sum_{\eta (\bmod q/h_2)} G(h, b, a, \overline{\eta}, q)$$
$$\times \sum_{n \le N} \xi\eta(n) \Lambda(n) e(n\lambda) + O(L^2),$$

where

$$\sum_{\xi(\bmod r/h_1)},$$

respectively

$$\sum_{\eta(\bmod q/h_2)}$$

are over the Dirichlet characters mod r/h_1 *respectively* q/h_2.

Proof This is Lemma 2 of [34]. □

We now decompose the sum in Lemma 3.3.6 in three partial sums.

Definition 3.3.7

$$S\left(\frac{a}{q}+\lambda, r, b\right) = S_0(a,q,\lambda,r,b) + S_1(a,q,\lambda,r,b) + S_2(a,q,\lambda,r,b)\,,$$

where S_0, S_1 and S_2, respectively, are the sums corresponding to

(0) $\xi = \xi^0 \left(\bmod \frac{r}{h_1}\right)$, $\eta = \eta^0 \left(\bmod \frac{q}{h_2}\right)$

(i) $\xi = \xi^0 \left(\bmod \frac{r}{h_1}\right)$, $\eta \neq \eta^0 \left(\bmod \frac{q}{h_2}\right)$

(ii) $\xi \neq \xi^0 \left(\bmod \frac{r}{h_1}\right)$,

respectively (ξ^0 resp. η^0 are the principal characters mod r/h_1 resp. mod q/h_2).

The asymptotics of the Major arcs contribution and thus also of (3.6) come from the sum S_0. To state the result, we need the following definitions:

Definition 3.3.8 Let q, r be positive integers and $(q,r) = h$. For $(a,q) = 1$ and $(b,r) = 1$, define

$$f(r,q,a,b) = \begin{cases} \frac{\mu(q/h)}{\phi(rq/h)}\, e\left(\frac{abt}{h}\right)\,, & \text{if } (q/h,h)=1,\, tq/h \equiv 1 (\bmod h) \\ 0\,, & \text{if } (q/h,h) > 1. \end{cases}$$

Then, we have:

Lemma 3.3.9

$$S_0 = f(r,q,a,b)\sum_{n\le N} e(n\lambda) + O(|\lambda| N^{3/2}\log^3 N) + O(N^{1/2}\log^3 N)\,.$$

Proof It holds

$$S_0 = \frac{1}{\phi(r/h_1)\phi(q/h_2)} G(h, b, a, \overline{\eta}, q) \sum_{n \leq N} \chi^0(n)\Lambda(n)e(n\lambda)$$

Since $\chi^0(n) = 1$ for all primes $n = p$ with $p \geq r$, it follows

$$\sum_{n \leq N} \chi^0(n)\Lambda(n)e(n\lambda) = \sum_{n \leq N} \Lambda(n)e(n\lambda) + O\left(N^{1/2} \log N\right) .$$

Thus,

$$\sum_{n \leq N} \Lambda(n)e(n\lambda) = \sum_{n \leq N} e(n\lambda) + \sum_{n \leq N} (\Lambda(n) - 1)e(n\lambda) .$$

By the Riemann Hypothesis (see [29, 30]), we have

$$\sum_{n \leq N} (\Lambda(n) - 1) = O(N^{1/2} \log^3 N)$$

and by partial summation:

$$\sum_{n \leq N} (\Lambda(n) - 1)e(n\lambda) = O(|\lambda| N^{3/2} \log^3 N) + O(N^{1/2} \log^3 N) .$$

The result now follows from Lemma 3.3.5. □

We now show that for $\frac{1}{2}R \leq r \leq 2R$ the contribution of S_1 and S_2 to the Major arcs contribution is negligible.

Definition 3.3.10 For $(b_1, b_2, b_3) \in \mathbb{Z}^3$ let

$$\int_0 = \int_{E_1(R)} S(\alpha, r, b_1)S(\alpha, r, b_2)S(\alpha, r, b_3)e(-N\alpha)d\alpha$$

and

$$\int_1 = \int_{E_1(R)} S_0(\alpha, r, b_1)S_0(\alpha, r, b_2)S_0(\alpha, r, b_3)e(-N\alpha)d\alpha .$$

Lemma 3.3.11 *Let*

$$\frac{1}{2}R \leq r \leq 2R, \ (b_1, b_2, b_3) \in \mathbb{Z}^3 .$$

Then

$$\int_0 = \int_1 + O(N^{11/4}R^{-2}) .$$

Proof We partition the sums $S(\frac{a}{q} + \lambda, r, b_i)$ according to Definition 3.3.7. The product

$$\prod_{i=1}^{3} S(\alpha, r, b_i)$$

becomes a sum of 27 products

$$S_{j_1}(\alpha, r, b_1)S_{j_2}(\alpha, r, b_2)S_{j_3}(\alpha, r, b_3)$$

with $j_i \in \{0, 1, 2\}$.

Assume that $(j_1, j_2, j_3) \neq (0, 0, 0)$. Without loss of generality, we may assume $j_1 \neq 0$. Then, the characters $\xi\eta$ appearing in the sums

$$\sum_{n \leq N} \xi\eta(n)\Lambda(n)e(n\lambda)$$

by Lemma 3.3.6 are nonprincipal and have conductor $\leq N$. Therefore, by the Generalized Riemann Hypothesis (see [29, 30]), we have the estimate

$$\sum_{n \leq N} \xi\eta(n)\Lambda(n) = O(N^{1/2}\log^3 N) + O(|\lambda|N^{3/2}\log^3 N) .$$

The other sums may be trivially estimated by

$$S_{j_i} = O(NR^{-1}) .$$

This proves Lemma 3.3.11. □

Lemma 3.3.12

$$\int_0 = \left(\frac{1}{2}N^2 + O(N^2L^{-C})\right) \sum_{q \leq P} \sum_{\substack{a=1 \\ (a,q)=1}}^{q} f(r, q, a, b_1) f(r, q, a, b_2) f(r, q, a, b_3) e\left(-\frac{aN}{q}\right) .$$

Proof From Lemmas 3.3.9 and 3.3.11, we obtain:

$$\int_0 = \sum_{q\le P}\sum_{a=1}^{q} f(r,q,a,b_1)f(r,q,a,b_2)f(r,q,a,b_3))e\left(-\frac{aN}{q}\right)$$
$$\times \int_{|\lambda|\le 1/qQ}\left(\sum_{n\le N} e(n\lambda)\right)^3 e(-N\lambda)d\lambda + O(N^{11/4}R^{-2}).$$

Using the estimate

$$\sum_{n\le N} e(n\lambda) \ll \min\left(N, \frac{1}{|\lambda|}\right)$$

one sees that the integral is

$$\int_{-1/2}^{1/2}\left(\sum_{n\le N} e(n\lambda)\right)^3 e(-N\lambda)d\lambda + O\left(\int_{1/qQ}^{1/2}\lambda^{-3}d\lambda\right) = \sum_{\substack{n_1+n_2+n_3=N\\ 1\le n_j\le N}} 1 + O((qQ)^2)$$
$$= \frac{1}{2}N^2 + O(N^2L^{-C}).$$

This completes the proof of Lemma 3.3.12. □

Lemma 3.3.13 *Let*

$$b_1+b_2+b_3 \equiv 0 (\mathrm{mod}\ r).$$

Then the singular series

$$\sigma(N;r) = \sum_{q=1}^{\infty}\sum_{a=1}^{q} f(r,q,a,b_1)f(r,q,a,b_2)f(r,q,a,b_3))e\left(-\frac{aN}{q}\right)$$

converges and has the value

$$\frac{1}{\phi^3(r)}\sum_{q=1}^{\infty}\frac{\mu(q/h)}{\phi^3(q/h)}\sum_{\substack{a=1\\(a,q)=1}}^{q} e\left(\frac{a(b_1+b_2+b_3)t}{h} - \frac{aN}{q}\right).$$

We also have (setting $C(r)=2$ if r is odd and $C(r)=8$ if r is even):

$$\sigma(N;r) = \frac{C(r)}{r^2}\prod_{p|r}\frac{p^3}{(p-1)^3+1}\prod_{\substack{p|N\\ p\nmid r}}\frac{(p-1)((p-1)^2+1)}{(p-1)^3+1}\prod_{p>2}\left(1+\frac{1}{(p-1)^3}\right)$$
$$\ge \frac{1}{r^2\log r}.$$

Proof This is due to Rademacher [46]. □

Lemma 3.3.14 *Let $\epsilon > 0$. Then we have*

$$\frac{1}{\phi^3(r)} \sum_{\substack{q>P \\ (q/h,h)=1}} \frac{\mu(q/h)}{\phi^3(q/h)} \sum_{\substack{a=1 \\ (a,q)=1}}^{q} e\left(\frac{a(b_1+b_2+b_3)t}{h} - \frac{aN}{q}\right) = O\left(\frac{r^{2+\epsilon}}{\phi^3(r)} P^{-1}\right).$$

Proof Using the estimates

$$\left| \sum_{\substack{a=1 \\ (a,q)=1}}^{q} e\left(\frac{a(b_1+b_2+b_3)t}{h} - \frac{aN}{q}\right) \right| \leq q,$$

$$d(m) \leq m^{\epsilon} \text{ for the divisor function}$$

and

$$\frac{m}{\phi(m)} = O(\log\log m),$$

we obtain

$$\begin{aligned}
&\frac{1}{\phi^3(r)} \sum_{\substack{q>P \\ (q/h,h)=1}} \frac{\mu(q/h)}{\phi^3(q/h)} \sum_{\substack{a=1 \\ (a,q)=1}}^{q} e\left(\frac{a(b_1+b_2+b_3)t}{h} - \frac{aN}{q}\right) \\
&\leq \frac{1}{\phi^3(r)} \sum_{\substack{q>P \\ (q/h,h)=1}} \frac{1}{\phi^3(q/h)} \cdot q \\
&\leq \frac{1}{\phi^3(r)} \sum_{h|r} \phi^3(h) \left(\sum_{\substack{q>P \\ h|q}} \frac{1}{q^2} \right) \log\log q \\
&\leq \frac{1}{\phi^3(r)} \sum_{h|r} \phi^3(h) \frac{1}{h^2} \sum_{\tilde{q}>P/h} \frac{1}{\tilde{q}^2} \log_2 q \\
&\leq \frac{1}{\phi^3(r)} P^{-1} \sum_{h|r} h^2 = O\left(\frac{r^{2+\epsilon}}{\phi^3(r)} P^{-1}\right).
\end{aligned}$$

□

Theorem 3.3.15 *Let*

$$R \leq N^{1/20}, \ b_1 + b_2 + b_3 = 0, \ \frac{R}{2} \leq R \leq 2R .$$

Then

$$\sum_{\substack{(n_1,n_2,n_3)\in\mathbb{N}^3 \\ n_i \equiv b_i (\mod r) \\ n_1+n_2+n_3=N}} \Lambda(n_1)\Lambda(n_2)\Lambda(n_3) = \frac{1}{2}\sigma(N;r)N^2(1+o(1))$$

as $N \to \infty$.

Proof This follows from (3.6), Lemma 3.3.2 and Lemmas 3.3.12, 3.3.13, 3.3.14. □

4 Conclusion

Theorem 3.1.3 now follows if we apply Theorem 3.3.15 with

$$b_1 = b_2 = u_0, \ b_3 = N - 2u_0, \ r = P^* .$$

Basic Steps of the Proof of Schnirelmann's Theorem

In this chapter, we provide an outline of the proof of Schnirelmann's theorem which states that there exists a positive integer q, such that every integer greater than 1 can be represented as the sum of at most q prime numbers. First, we define Schnirelmann's density, which plays an integral role in the proof of Schnirelmann's theorem, and afterwards we shall mention the basic steps of that proof (cf. [28, 58]).

Definition 4.0.1 Let $\mathbb{D}$ denote a set of distinct integers m, where $m \in \mathbb{N} \cup \{0\}$. Let $A(n)$ be the number of nonzero elements of $\mathbb{D}$ which do not exceed the positive integer n, that is,

$$A(n) = \sum_{\substack{m \in \mathbb{D} \\ 1 \leq m \leq n}} 1 .$$

If there exists a real number c, $c > 0$, such that

$$A(n) \geq cn , \tag{D1}$$

for every $n \in \mathbb{N}$, then we say that $\mathbb{D}$ is a **positive density set**. Moreover, the supremum of c's, for which (D1) holds true, is called **the density of the set** $\mathbb{D}$.

Remark 4.0.2 Since $1 \geq A(1) \geq c > 0$, it follows that $A(1) = 1$. Hence, $1 \in \mathbb{D}$.

Now that, we have established the definition of a positive density set, and we shall define the sum of two positive density sets.

Definition 4.0.3 Let $\mathbb{D}_1$, $\mathbb{D}_2$ be two positive density sets. Then, the set

$$\mathbb{S} = \{m_1 + m_2 \,|\, m_1 \in \mathbb{D}_1 \text{ and } m_2 \in \mathbb{D}_2\}$$

M.T. Rassias, *Goldbach's Problem*, DOI 10.1007/978-3-319-57914-6_4

is called **the sum of the positive density sets** $\mathbb{D}_1$, $\mathbb{D}_2$, and we write

$$\mathbb{S} = \mathbb{D}_1 + \mathbb{D}_2 \, .$$

The following result provides a lower estimate for the density s of $\mathbb{S}$ in terms of the densities c_1, c_2 of $\mathbb{D}_1$, $\mathbb{D}_2$, respectively.

Theorem 4.0.4 *Let $\mathbb{D}_1$, $\mathbb{D}_2$ be two positive density sets with densities c_1, c_2, respectively. Then, if $0 \in \mathbb{D}_1$ and $\mathbb{S} = \mathbb{D}_1 + \mathbb{D}_2$, for the density s of $\mathbb{S}$, we have*

$$s \geq c_1 + c_2 - c_1 c_2 \, .$$

Note The symbols $\mathbb{D}_1, \mathbb{D}_2, \mathbb{S}, c_1, c_2$ and s will have the same meaning throughout this section and for that reason we shall not define them again.

Proof Let m_{1i}, m_{2i} denote elements of $\mathbb{D}_1$, $\mathbb{D}_2$, respectively, where

$$m_{11} < m_{12} < \ldots < m_{1A_1(n)} \, ,$$

$$m_{21} < m_{22} < \ldots < m_{2A_2(n)} \, ,$$

and

$$A_j(n) = \sum_{\substack{m_{ji} \in \mathbb{D}_j \\ 1 \leq m_{ji} \leq n}} 1 \, , \quad j = 1, \; 2 \, .$$

Since, we know that $c_2 > 0$, it is evident that

$$A_2(n) \geq c_2 n \, ,$$

for every $n \in \mathbb{N}$. Thus,

$$A_2(1) \geq c_2 > 0$$

or

$$\sum_{1 \leq m_{2i} \leq n} 1 \geq c_2 > 0 \, .$$

Hence, there exists $m_{2i} \in \mathbb{D}_2$. Therefore, if we set $m_{21} = 1$, then due to the fact that $0 \in \mathbb{D}_1$, it follows that

$$m_{21}, \; m_{22}, \; \ldots \; , \; m_{2A_2(n)} \in \mathbb{S} \, .$$

Thus, for a start we obtain that

$$S(n) \geq A_2(n) , \tag{1}$$

where

$$S(n) = \sum_{\substack{k \in \mathbb{S} \\ 1 \leq k \leq n}} 1 .$$

We shall now examine the numbers

$$m_{1i} + m_{2j} ,$$

for a fixed i and for $j = 1, 2, \ldots, A_2(n) - 1$, where

$$1 \leq m_{1i} \leq m_{2(j+1)} - m_{2j} - 1 .$$

We obviously have

$$m_{1i} + m_{2j} \geq 1 + m_{2j} . \tag{2}$$

In addition,

$$\begin{aligned} m_{1i} + m_{2j} &\leq (m_{2(j+1)} - m_{2j} - 1) + m_{2j} \\ &= m_{2(j+1)} - 1 \\ &\leq m_{2A_2(n)} - 1 . \end{aligned}$$

Thus, it is clear that

$$m_{1i} + m_{2j} \leq n - 1 . \tag{3}$$

Hence, from (2), (3) we obtain that

$$1 + m_{2j} \leq m_{1i} + m_{2j} \leq n - 1 ,$$

for a fixed i and for $j = 1, 2, \ldots, A_2(n) - 1$.
Therefore, the natural numbers

$$m_{1i} + m_{2j}$$

are pairwise distinct and less than n.
Since the numbers $m_{1i} + m_{2j}$ and m_{2j} are also pairwise distinct, from the above arguments and (1), it follows that

$$S(n) \geq A_2(n) + A_1(n - m_{2A_2(n)}) + \sum_{j=1}^{A_2(n)-1} A_1(m_{2(j+1)} - m_{2j} - 1) .$$

However, since $A_1(n) \geq c_1 n$ for every $n \in \mathbb{N}$, it yields

$$S(n) \geq A_2(n) + c_1(n - m_{2A_2(n)}) + \sum_{j=1}^{A_2(n)-1} c_1(m_{2(j+1)} - m_{2j} - 1) . \quad (4)$$

But

$$(m_{22} - m_{21} - 1) + (m_{23} - m_{23} - 1) + \cdots + (m_{2A_2(n)} - m_{2A_2(n)-1} - 1)$$
$$= m_{2A_2(n)} - m_{21} - (A_2(n) - 1) .$$

Hence, from (4) it follows that

$$S(n) \geq A_2(n) + c_1(n - m_{2A_2(n)}) + c_1 \left(m_{2A_2(n)} - m_{21} - (A_2(n) - 1)\right)$$
$$= A_2(n) + c_1(n - m_{21} + 1 - A_2(n))$$

and since $m_{21} = 1$, we get

$$S(n) \geq A_2(n) + c_1(n - A_2(n))$$
$$= c_1 n + (1 - c_1)A_2(n) .$$

However, since $A_1(n) \leq n$, it is evident that $c_1 \leq 1$. Thus, $1 - c_1 \geq 0$ and due to the fact that $A_2(n) \geq c_2 n$, we get

$$S(n) \geq c_1 n + (1 - c_1)c_2 n$$
$$= n(c_1 + c_2 - c_1 c_2) .$$

By the above relation, it is clear that for the density s of $\mathbb{S}$, it holds

$$s \geq c_1 + c_2 - c_1 c_2 . \qquad \square$$

We shall now proceed to the proof of a theorem related to the sum of positive density sets.

Theorem 4.0.5 *If* $0 \in \mathbb{D}_1$ *and* $c_1 + c_2 \geq 1$, *then it holds*

$$s = 1 .$$

(i.e. the sum $\mathbb{S}$ of the positive density sets $\mathbb{D}_1$, $\mathbb{D}_2$ is identical to the set of natural numbers $\mathbb{N}$).

Proof We shall prove this theorem by contradiction. Let us suppose that $s \neq 1$, which means that $s < 1$. Then, it is clear that $\mathbb{S} \neq \mathbb{N}$ and thus, there exist positive integers n, such that $n \notin \mathbb{S}$. Let n_0 be the least positive integer, such that $n_0 \notin \mathbb{S}$. Then,

$$n_0 > 1 \text{ and } n_0 \notin \mathbb{D}_2 .$$

These two statements hold true due to the following reasoning.
We have already shown in the proof of Theorem 4.0.4 that since $c_2 > 0$, there exists $m_2 = 1 \in \mathbb{D}_2$. However, since $0 \in \mathbb{D}_1$, it is evident that $1 \in \mathbb{S}$ and for that reason $n_0 > 1$. Furthermore, if we had $n_0 \in \mathbb{D}_2$, then due to the fact that $0 \in \mathbb{D}_1$ it would follow that $n_0 \in \mathbb{S}$, which is a contradiction. We now observe that for every $m_1, m_2 \in \mathbb{N}$, where $m_1 \in \mathbb{D}_1, m_2 \in \mathbb{D}_2$, we have

$$n_0 \neq m_1 + m_2 . \tag{1}$$

Therefore, by the use of the above basic property, we can construct two disjoint sets B, C of positive integers as follows:

$$B = \{m_1 \,|\, m_1 \in \mathbb{D}_1 \text{ and } 1 \leq m_1 \leq n_0 - 1\}$$

and

$$C = \{n_0 - m_2 \,|\, m_2 \in \mathbb{D}_2 \text{ and } 1 \leq m_2 \leq n_0 - 1\} .$$

The above sets are clearly disjoint. Indeed, if there is an element $k \in B \cap C$, then we have

$$m_1 = k = n_0 - m_2 ,$$

which is a contradiction, due to (1).
Hence, for the cardinality of the set $B \cup C$, we obtain

$$\text{Card}\,(B \cup C) \leq n_0 - 1 . \tag{2}$$

However,

$$\text{Card}\,(B \cup C) = A_1(n_0 - 1) + A_2(n_0 - 1) ,$$

where

$$A_i(n) = \sum_{1 \leq m_i \leq n} 1 , \quad m_i \in \mathbb{D}_i , \quad i = 1, 2 .$$

But, since $n_0 \notin \mathbb{D}_2$, it is evident that

$$A_2(n_0 - 1) = A_2(n_0) .$$

Therefore,

$$\begin{aligned} \text{Card}\,(B \cup C) &= A_1(n_0 - 1) + A_2(n_0) \\ &\geq c_1(n_0 - 1) + c_2 n_0 \\ &> c_1(n_0 - 1) + c_2(n_0 - 1) \\ &= (c_1 + c_2)(n_0 - 1) \end{aligned}$$

and by the hypothesis that $c_1 + c_2 \geq 1$, we get

$$\text{Card}\,(B \cup C) > n_0 - 1 ,$$

which is a contradiction, due to (2). Thus, the assumption that $s \neq 1$ leads to a contradiction and consequently $s = 1$. This completes the proof of the theorem. □

Theorem 4.0.6 *Let the positive density set $\mathbb{D}$ have density c, $c < 1$. Let*

$$k = 2\left(\left\lfloor -\frac{\log 2}{\log(1-c)} \right\rfloor + 1\right) .$$

(a) If $0 \in \mathbb{D}$, then every natural number can be represented as the sum of k elements of $\mathbb{D}$.
(b) If $0 \notin \mathbb{D}$, then every natural number can be represented as the sum of at most k elements of $\mathbb{D}$.

Proof (a) Since we are trying to prove a theorem of representation as a sum of k elements of $\mathbb{D}$, it is evident that we must construct a set of the form

$$\underbrace{\mathbb{D} + \mathbb{D} + \cdots + \mathbb{D}}_{m \text{ summands}} .$$

We denote the above set by $\mathbb{S}_m$. Our goal is to prove that the set $\mathbb{S}_k$ is identical to the set $\mathbb{N}$ of natural numbers. Equivalently, it suffices to prove that the density s_k of $\mathbb{S}_k$ is equal to 1. However, by the previous theorem we observe that if

$$\mathbb{S}_k = \mathbb{S}_{k/2} + \mathbb{S}_{k/2} ,$$

[1]with $s_{k/2} + s_{k/2} \geq 1$, then

$$s_k = 1\,.$$

Hence, we shall try to find a lower bound for the density s_m of the set $\mathbb{S}_m$. More specifically, we will prove by induction that

$$s_m \geq 1 - (1-c)^m\,, \tag{1}$$

for every natural number $m \in \mathbb{N}$.
For $m = 1$, it is clear that (1) holds true, since $s_1 = c$. Now, we assume that

$$s_{m-1} \geq 1 - (1-c)^{m-1} \tag{2}$$

is true. Then, since

$$\mathbb{S}_m = \mathbb{D} + \mathbb{S}_{m-1}\,,$$

by Theorem 4.0.4, it follows that

$$s_m \geq c + s_{m-1} - cs_{m-1}$$

and by (2), we get

$$\begin{aligned} s_m &\geq c + (1-c)\left(1 - (1-c)^{m-1}\right) \\ &= 1 - (1-c)^m\,. \end{aligned}$$

Hence, (1) is true for every $m \in \mathbb{N}$.
Therefore,

$$\begin{aligned} s_{k/2} + s_{k/2} = 2s_{k/2} &\geq 2\left(1 - (1-c)^{k/2}\right) \\ &= 2\left(1 - (1-c)^{\lfloor -\log 2/\log(1-c)+1 \rfloor}\right) \\ &\geq 2\left(1 - (1-c)^{-\log 2/\log(1-c)}\right) \\ &= 2\left(1 - e^{-\log 2}\right) = 2\left(1 - \frac{1}{2}\right) \\ &= 1\,. \end{aligned}$$

This completes the proof of (a).
(b) If $0 \notin \mathbb{D}$, then we consider the set

$$\mathbb{D}^* = \mathbb{D} \cup \{0\}\,.$$

---

[1]Note that $k/2$ is a natural number, since k is an even positive integer.

Hence, by (a) it follows that every natural number can be represented as the sum of k elements of $\mathbb{D}^*$. However, since some of those elements might be equal to 0, it is evident that in this case every natural number can be represented as the sum of at most k elements of $\mathbb{D}$. □

We shall now proceed to the presentation of the basic ideas behind the proof of Schnirelmann's theorem.

Theorem 4.0.7 (SCHNIRELMANN'S THEOREM) *There exists a positive integer q, such that every integer greater than 1 can be represented as the sum of at most q prime numbers.*

Basic steps of the proof

The first key step of the proof is to consider the set

$$\mathbb{H} = \{p_1 + p_2 \mid p_1,\ p_2 \text{ are prime numbers}\} \cup \{1\} .$$

Now, let us make the assumption that $\mathbb{H}$ is a positive density set. If that is true, then by the previous remark, Theorem 4.0.6 and the fact that $0 \notin \mathbb{H}$ the following Lemma holds:

Lemma 4.0.8 *Assume that $\mathbb{H}$ has positive density. Then, every natural number can be represented as the sum of at most k elements of the set $\mathbb{H}$.*

Lemma 4.0.9 *Every integer $r \geq 2$ is the sum of at most $r/2$ prime numbers.*

Proof If r is even, we can write

$$r = \underbrace{2 + \cdots + 2}_{r/2 \text{ summands}} .$$

If r is odd, we can write

$$r = 3 + \underbrace{2 + \cdots + 2}_{(r-3)/2 \text{ summands}} .$$

Therefore, Lemma 4.0.9 follows. □

Lemma 4.0.10 *Assume that $\mathbb{H}$ has positive density. Then, each integer $m \geq 3$ can be written as the sum of at most $2k$ primes.*

Proof By Lemma 4.0.8, m can be represented as a sum of at most $2k$ positive integers, each of which is either one or a prime number. The same is true for $m - 1$.

Let,

$$m = \underbrace{1 + \cdots + 1}_{r \text{ summands}} + \sum_{i=1}^{l-r} p_i \,, \tag{1}$$

with $l \leq 2k$ and p_i primes.

Case 1. Let $r \geq 2$. Then by Lemma 4.0.8, the number r can be written as a sum of at most $r/2$ primes, which by (1) gives m as a sum of at most $2k$ primes.

Case 2. Let $r = 1$. By Lemma 4.0.8, the number $m - 1$ can be written as a sum of at most $2k$ positive integers, each of which is either 1 or a prime number:

$$m - 1 = \underbrace{1 + \cdots + 1}_{r' \text{ summands}} + \sum_{i=1}^{l'-r'} q_i \,, \tag{2}$$

with $l' \leq 2k$ and q_i primes.

By Lemma 4.0.9, we can write $r' + 1$ as a sum of at most $(r' + 1)/2$ primes, which by (2) gives m as a sum of at most $2k$ primes. Therefore, the proof of Lemma 4.0.10 is complete. □

Therefore, by the above arguments it is clear that in order to prove Schnirelmann's theorem, it suffices to prove that $\mathbb{H}$ is a positive density set. At this point, we shall define **the function of multiplicity of the elements** of $\mathbb{H}$ and then prove a simple lemma, which will be useful later.

Definition 4.0.11 Let $\varepsilon_{\mathbb{H}}(m)$ denote **the multiplicity of the element** m in $\mathbb{H}$. Then, we define $\varepsilon_{\mathbb{H}}(1) = 1$. For $m > 1$, we have

$$\varepsilon_{\mathbb{H}}(m) = \sum_{p_1 + p_2 = m} 1 \,,$$

where the sum extends over all pairs of prime numbers p_1, p_2, such that

$$p_1 + p_2 = m \,.$$

Lemma 4.0.12 *Let $\mathbb{G}$ be a sequence of nonnegative integers. If $m \in \mathbb{G}$ and for any natural number n it holds*

$$\left(\sum_{1 \leq m \leq n} \varepsilon(m) \right)^2 \geq n c_0 \sum_{1 \leq m \leq n} \varepsilon^2(m) \,, \tag{L1}$$

where $\varepsilon(m)$ denotes the multiplicity of m in $\mathbb{G}$ and c_0 is a positive constant, then $\mathbb{G}$ is a positive density set.

Proof By the Cauchy-Schwarz-Buniakowsky inequality, we have

$$\sum_{1\le m\le n} \varepsilon^2(m) \sum_{1\le m\le n} 1^2 \ge \left(\sum_{1\le m\le n} \varepsilon(m)\right)^2$$

or

$$\sum_{1\le m\le n} 1 \ge \left(\sum_{1\le m\le n} \varepsilon(m)\right)^2 \left(\sum_{1\le m\le n} \varepsilon^2(m)\right)^{-1}.$$

Thus, by the hypothesis we obtain

$$\sum_{1\le m\le n} 1 \ge nc_0\,,$$

for every positive integer n. Therefore, it is clear that $\mathbb{G}$ is a positive density set. □

By the above lemma, it is evident that it suffices to prove that there exists a positive constant c_0', such that inequality (L1) holds true with respect to $\varepsilon_{\mathbb{H}}(m)$, for every natural number n.

Theorem 4.0.13 *For any positive integer n with $n \ge 2$, there exists a positive constant c_1, such that*

$$\left(\sum_{1\le m\le n} \varepsilon_{\mathbb{H}}(m)\right)^2 \ge c_1 \frac{n^4}{\log^4 n}\,.$$

Proof We have

$$\begin{aligned}
\sum_{1\le m\le n} \varepsilon_{\mathbb{H}}(m) &= \varepsilon_{\mathbb{H}}(1) + \sum_{2\le m\le n} \varepsilon_{\mathbb{H}}(m) \\
&= 1 + \sum_{4\le m\le n} \varepsilon_{\mathbb{H}}(m) \quad (\text{since } \varepsilon_{\mathbb{H}}(2) = \varepsilon_{\mathbb{H}}(3) = 0) \\
&= 1 + \sum_{4\le m\le n} \sum_{p_1+p_2=m} 1 \\
&= 1 + \sum_{4\le p_1+p_2\le n} 1
\end{aligned}$$

$$\geq \sum_{p_1 \leq n/2} 1 \cdot \sum_{p_2 \leq n/2} 1$$
$$= \pi\left(\frac{n}{2}\right) \cdot \pi\left(\frac{n}{2}\right) .$$

By Chebyshev's inequality, we know that

$$\pi(n) > \frac{1}{6} \frac{n}{\log n} ,$$

for any positive integer n, with $n \geq 2$.
Therefore, it is obvious that there exists a positive integer c, such that

$$\sum_{1 \leq m \leq n} \varepsilon_{\mathbb{H}}(m) \geq c \frac{n^2}{\log^2 n} .$$

Thus, for $c_1 = c^2$ we get

$$\left(\sum_{1 \leq m \leq n} \varepsilon_{\mathbb{H}}(m) \right)^2 \geq c_1 \frac{n^4}{\log^4 n} .$$

This completes the proof of Theorem 4.0.13. □

The final step to prove Schnirelmann's theorem is to find a suitable upper bound for the sum

$$\sum_{1 \leq m \leq n} \varepsilon_{\mathbb{H}}^2(m) .$$

This upper bound is provided by the following theorem.[2]

Theorem 4.0.14 *For every positive integer n, with $n \geq 2$, there exists a positive constant c_2, such that*

$$\sum_{1 \leq m \leq n} \varepsilon_{\mathbb{H}}^2(m) \leq c_2 \frac{n^3}{\log^4 n} .$$

By Theorems 4.0.14 and 4.0.13, it follows that there exists a positive constant $c_0 = c_1/c_2$, such that inequality (L1) of Lemma 4.0.12 holds true.
Hence, $\mathbb{H}$ is a positive density set and this completes the proof of Schnirelmann's theorem.

---

[2]For a proof see [28].

Appendix

4.1 A Sketch of H. Helfgott's Proof of Goldbach's Ternary Conjecture—By Olivier Ramaré

In 2014, Harald Helfgott published a preprint containing a proof that every odd integer strictly larger than 5 is a sum of three primes. The proof is still to be verified, but it is likely to be correct. The fact that this proof has now some slack goes in this direction. The aim of this appendix is to sketch it and to try to underline the major argument. Concerning the latest approximation of this result, we should mention that Ming-Chit Liu and Tianze Wang had shown in 2002 in [35] that every odd integer larger than exp(3100) is indeed a sum of three primes and that Terence Tao had shown in 2014 in [60] that every odd integer larger than 5 is a sum of at most five primes.

We stay close to the notation of Helfgott's proof to help the reader follow it, but we simplify some expressions and parameters. The proof under examination follows the path drawn by Hardy and Littlewood in [19] and Vinogradov [66] and considers, when $N \geq 10^{27}$, the quantity

$$R_3(N) = \sum_{p_1+p_2+p_3=N} \eta_+\left(\frac{p_1}{x}\right) \log p_1 \, \eta_+\left(\frac{p_2}{x}\right) \log p_2 \, \eta_*\left(\frac{p_3}{x}\right) \log p_3 \, , \quad (4.1)$$

where x is a size parameter close to $N/2$, and η_+ and η_* are two C^∞ nonnegative smooth functions whose forms are chosen so as to optimize the estimates. Let us say that $\eta_+(t)$ is close to

$$t^3 \max(0,\ 2-t)^3 e^{t-(t^2+1)/2}$$

and bounded above by 1.08, while $\eta_*(t)$ also contains a $e^{-t^2/2}$-part but is more complicated and is bounded above by 1.42. Here is the main theorem.

M.T. Rassias, *Goldbach's Problem*, DOI 10.1007/978-3-319-57914-6

Theorem 4.1.1 *If N is odd and larger than 10^{27}, we have*

$$R_3(N) \geq \frac{N^2}{5000}.$$

From the above theorem, we infer that every odd integer $\geq 10^{27}$ is a sum of three primes. Computations run by Harald Helfgott and David Platt in [25] show that every odd integer within

$$[\,7,\ 8.875 \cdot 10^{30}\,]$$

is a sum of three odd primes, concluding the proof. The paper [31] by Habiba Kadiri and Alyssa Lumley on explicit short intervals containing primes could also lead to the same result.

The base line of the argument of Theorem 4.1.1 is the proof of Vinogradov that the reader will find in [64, Chap. 4] and in the present book, which is devoted to a step-by-step proof of this theorem. We introduce, for an arbitrary function η, the trigonometric polynomial

$$S_\eta(\alpha, x) = \sum_{p \geq 3} \eta(p/x) \log p\ e(\alpha x)$$

and use the easily checked identity

$$R_3(N) = \int_0^1 S_{\eta_+}(x, \alpha)^2 S_{\eta_*}(x, \alpha) e(-N\alpha) d\alpha. \tag{4.2}$$

For a positive integer q, we define the following subset of $\mathbb{R}/\mathbb{Z}$:

$$\mathfrak{M}_{q,\delta,r} = \bigcup_{\substack{1 \leq a \leq q \\ (a,q)=1}} \left[\frac{a}{q} - \frac{\delta r}{2qx}, \frac{a}{q} + \frac{\delta r}{2qx}\right]. \tag{4.3}$$

We set $r_0 = 150\,000$ and $Q = (3/4)x^{2/3}$. We split the interval $[0, 1]$ in three subsets.

Majors Arcs
When

$$\alpha \in \bigcup_{q \leq r_0} \mathfrak{M}_{q,8,r_0},$$

the value of $S_\eta(\alpha, x)$ in (4.2) is controlled by the distribution of primes in progressions modulo q. From an explicit viewpoint and due to [57] and [39], we know that the clear path to such information follows by checking a partial Riemann Hypothesis, that is, by showing that the L—functions attached to Dirichlet characters modulo q have no zero $\rho = \beta + i\gamma$ inside the set

$$\{1/2 < \beta < 1,\ |\gamma| \leq T_q\}$$

for some fixed T_q. Such computations are highly complex, and until recently, we were only able to roughly use $q \leq 60$ and $T_q = 10\,000$. David Platt's new algorithm from [45] changed all that, and we can now use values of q up to 300 000 with

$$T_q = \frac{10^8}{q}.$$

Rather than splitting the sum

$$S_\eta\left(\frac{a}{q} + u,\ x\right) = \sum_p \eta\left(\frac{p}{x}\right) e(pu) e\left(\frac{ap}{q}\right)$$

according to the class of p modulo q, Helfgott goes back to the initial method, expresses the additive character $n \mapsto e(an/q)$ restricted to the multiplicative group $(\mathbb{Z}/q\mathbb{Z})^*$ in terms of Dirichlet characters and uses an explicit formula. This means also controlling the Mellin transform of the used gaussian smoothing, a task in which he invests a lot of effort with a successful outcome. A main term of size of order x^2 is extracted in this manner.

Minor Arcs

We now present the core of one of the main steps.

Lemma 4.1.2 *If*

$$\alpha = \frac{a}{q} + \frac{\tau}{qQ}$$

with $|\tau| \leq 1$, *we set* $\delta^* = 2 + |\tau x|$. *For* $q \leq x^{1/3}$, *we have*

$$S_{\eta_*}(\alpha, x) \ll \frac{x \log(\delta^* q)}{\sqrt{\delta^* \varphi(q)}}.$$

For $x^{1/3} \leq q \leq Q$, *we have*

$$S_{\eta_*}(\alpha, x) \ll x^{5/6} (\log x)^{3/2}.$$

The implied constants are small. The second part is rather classical, but the first one contains two essential novelties: when τ is small, the bound tends to zero as q goes to infinity and is not contaminated by any multiplicative factor like $\log x$, while the constant remains small. It is the first time both properties are kept, see for instance [8] and [49]. Furthermore, when τ is large, one gets *additional* saving, while earlier treatments all handled this factor as a perturbation of the case $\tau = 0$ and thus became worse when τ increased. Let us only mention that the method uses the bilinear form of [63] with $UV = x/\sqrt{q}$ when q is small. This lemma requires also an estimate of the form

$$\sum_{n\le x}\left(\sum_{d|n}\mu(d)1_{d\le z}\right)^2 \ll x\,,$$

where z is some power of x. With an unspecified constant, this is due to [11]. The treatment Helfgott devises relies on [51], which itself relies on [1] and [50].

The corresponding treatment of (4.2) follows the argument already contained in [66]:

$$\begin{aligned}\int_{\alpha\in\mathfrak{m}} S_{\eta_+}(x,\alpha)^2 S_{\eta_*}(x,\alpha)e(-N\alpha)d\alpha &\ll \int_0^1 |S_{\eta_+}(x,\alpha)|^2 d\alpha \max_{\alpha\in\mathfrak{m}} |S_{\eta_*}(x,\alpha)| \\ &\ll x^2\log x\cdot \max_{\alpha\in\mathfrak{m}} |S_{\eta_*}(x,\alpha)|/x \quad (4.4)\end{aligned}$$

for a subset $\mathfrak{m}$ that we call Minor arcs. If we select for $\mathfrak{m}$ the complement of the Major arcs, we can only ensure that

$$\max_{\alpha\in\mathfrak{m}}\frac{|S_{\eta_*}(x,\alpha)|}{x}$$

is smaller than a constant, but cannot recover the loss of $\log x$. Tao in [60] uses the fact that by [35], one can assume that $\log x \le 3100$. This bound is too large here. Helfgott introduces another subset of the circle: the intermediate arcs.

Intermediate Arcs

The key to the intermediate arcs is a lemma of the following form.

Lemma 4.1.3 *Let (φ_n) be a sequence in $\ell_1\cap\ell_2$ such that $\varphi_n = 0$ when n has a prime factor which is at most $\sqrt{x}$. Let $Q_0 \ge 10^5$ be such that*

$$Q_0 \le \frac{Q}{20000}\ \text{with}\ Q = \frac{\sqrt{x}}{4},\ \text{and}\ Q_0 \le Q^{3/5}\,.$$

We have

$$\sum_{q \le Q_0} \int_{\mathfrak{M}_{q,8,Q_0}} \Big| \sum_{n \ge 1} \varphi_n e(n\alpha) \Big|^2 d\alpha \le \frac{\log Q_0 + 1.36}{\log Q + 1.33} \sum_n |\varphi_n|^2$$

This is essentially a circle method version of [52, Theorem 5] or more explicitly of [48, Theorem 5.3]. As in (4.4), the norm $\sum_n |\varphi_n|^2$ in our case is of size $x \log x$, potentially loosing a $\log x$, but it is recovered by $1/\log Q$. Applying summation by parts and using Lemma 4.1.2, we get something of the type

$$\int_{\alpha \in \mathfrak{m}'} S_{\eta_+}(x, \alpha)^2 S_{\eta_*}(x, \alpha) e(-N\alpha) d\alpha \ll x \cdot \max_{\alpha \in \mathfrak{m}'} (\log q) |S_{\eta_*}(x, \alpha)|/x$$

where

$$\alpha = \frac{a}{q} + \frac{\tau}{qQ}$$

and

$$\mathfrak{m}' = \bigcup_{r_0 < q \le r_1} \mathfrak{M}_{q,8,q}$$

with r_1 being approximately $x^{4/15}/8$.

This is not quite enough in order to conclude: we need the arc $\mathfrak{M}_{q,8,r_1}$ rather than the arc $\mathfrak{M}_{q,8,q}$, but it is another place where the enhanced effect of δ^* in Lemma 4.1.2 features!

Note There are several other interesting techniques in this proof that cannot fit in this short sketch.

Olivier Ramaré
April, 2015

4.2 Some Biographical Remarks

In this section, we present briefly the biographies of some mathematicians who, over the years, proved major results related to Goldbach's Weak and Strong Conjectures, namely the biographies of C. Goldbach, G. H. Hardy, J. E. Littlewood, I. M. Vinogradov, L. G. Schnirelmann, J. Chen and H. A. Helfgott.

4.2.1 Christian Goldbach

Christian Goldbach was born in Königsberg, the capital of East Prussia, on the March 18 1690.
He studied at the Royal Albertus University at Königsberg. When he was just 20 years old, he decided to begin an educational voyage around Europe, which lasted for 14 years. During this voyage, he had the opportunity to meet some very important mathematicians of his time. More specifically, in 1711 he met Gottfried Leibniz in Leipzig; in 1712, he met Nicolaus I Bernoulli and Abraham de Moivre in London; and in 1721, he met Nicolaus II Bernoulli in Venice.
In 1720, Goldbach published his first paper entitled "Specimen methodi ad summas serierum", in Acta Eruditorum, the first scientific journal of the German lands. Three years later, he started a correspondence with Daniel Bernoulli, after the suggestion of D. Bernoulli's brother N. II Bernoulli. This correspondence continued until the end of the 1720s.
Meanwhile, in 1724, Goldbach decided to put an end to his long educational voyage and in 1725 he moved to St. Petersburg, where he started working in the newly opened St. Petersburg Academy of Sciences. He worked in the Academy until January, 1728, when he was asked to become the tutor of the very young Russian emperor at the time Peter II, and thus moved to Moscow. The next year, he started an important correspondence with Leonhard Euler, which went on for over three decades. During their correspondence, apart from the formulation of his famous conjectures in 1742, Goldbach discussed with Euler several topics of Number Theory, such as perfect numbers, Mersenne and Fermat numbers, Waring's problem.
Two years after the death of Peter II in 1730, Goldbach left Moscow and returned to St. Petersburg, where he started working again in the St. Petersburg Academy. However, in 1742, he obtained a high-rank position at the Russian Ministry of Foreign Affairs and thus stopped being involved with the Academy.
C. Goldbach died on the November 20 1764, in Moscow at the age of 74.

4.2.2 Godfrey Harold Hardy

Godfrey Harold Hardy was born on the February 7 1877, in the village Cranleigh, Surrey, England. His parents were known to be both intelligent and mathematically skilled, even though they never had the opportunity to receive proper university education.

Hardy's intimate relation with numbers started in the astonishingly young age of two, when it is said that he was able to count and write numbers up to millions. His inclination to number theory was also apparent, since as a very young child he enjoyed factoring the numbers of the hymns when taken to church (cf. [32]). He attended Cranleigh school until the age of 12, when he was bestowed a scholarship to Winchester College, one of the best schools in the country for mathematical education. In 1896, he entered the Trinity College of the University of Cambridge after being offered a scholarship. At Cambridge, Hardy sat for the Mathematical Tripos examination of 1898, the oldest Tripos examined by Cambridge and one of the most demanding mathematical examinations in the world until today. He managed to place fourth wrangler even though he prepared for only two years. Interestingly, the fact that he did not place first among his fellow candidates seemed to annoy him.

In 1900, he was elected a fellow of Trinity College and a year later he was bestowed jointly with James Hopwood Jeans the Smith's Prize by the University of Cambridge. In 1903, he graduated from Cambridge, and three years later, he was offered a lecturer position. In 1910, he was elected as a Fellow of the Royal Society.

In 1911, his long-lasting collaboration with J. E. Littlewood started, which was meant to be one of the most fruitful and celebrated collaborations in English mathematics. It was said as a joke at the time that there were three great English mathematicians at that period: Hardy, Littlewood and Hardy-Littlewood. In a lecture given by Harald Bohr on the occasion of his sixtieth birthday in 1947, among other things he mentioned four *axioms* that Hardy and Littlewood considered as rules for their mutual collaboration. These rules would secure some kind of independence and freedom between them (cf. [4], pp. 8–11). The first axiom stated that it was of no importance whatsoever whether what was written from one to the other, during their correspondance, was right or wrong. The second axiom, mainly focusing on the preservation of some kind of freedom, stated that when one received a letter from the other he was under no obligation to answer it or even study its content. According to the third axiom, although it did not really matter if they both thought about the same detail, it was preferable that they didn't. The last among the axioms stated that the extent to which one contributed to a joint paper was completely indifferent.

In 1913, Hardy received the first letter by S. Ramanujan and, after realizing the great mathematical talent of the sender, arranged to invite the young Indian autodidact at Cambridge. This lead to a major collaboration that influenced both Hardy's mathematical career and personal life. He was impressed by Ramanujan's genius to such a degree that he considered him to be of the calibre of Euler and Gauss. It is also worth mentioning that in an interview that Hardy gave to Paul Erdős, he described the discovery of Ramanujan as his most important contribution to Mathematics.

In 1918, Hardy and Ramanujan [21] introduced for the first time the Circle Method in a paper concerning partitions. However, Hardy and Littlewood developed that method so that it could be used to connect exponential sums with general problems of additive number theory. More specifically, in 1923, Hardy and Littlewood used the Circle Method and a modified form of the Riemann Hypothesis to prove that there exists a natural number N, such that every odd integer $n \geq N$ can be expressed as the sum of three prime numbers (cf. [19], [20]).

Notwithstanding the significance of Hardy's collaborations with Littlewood and Ramanujan, he also worked with several other mathematicians among which were E. Landau, E. C. Titchmarsh, G. Pólya and E. M. Wright.

Since Hardy was unhappy in Cambridge, in 1919 he grasped the opportunity to move to Oxford after being offered the position of Savilian Professor. He considered his research years there as some of the most fruitful of his entire career. Additionally, during those years he established in the University of Oxford a school of Analysis with E. C. Titchmarsh as his first student.

One year after moving to Oxford, he was awarded the Royal Medal of the Royal Society of London (RSL). In 1928, he moved for a year to Princeton in an exchange with Oswald Veblen who moved to Oxford, respectively. In 1929, he was also awarded the De Morgan Medal of the RSL.

After the retirement of Ernest W. Hobson in 1931, Hardy moved back to Cambridge and held the Sadleirian Chair in Pure Mathematics. In 1940 and 1947, he was bestowed the Sylvester and the Copley Medals of the RSL, respectively.

Other than Mathematics, Hardy is known for his peculiar eccentricities and for his love for cricket. Among others, Hardy was a raving atheist. It is amusing to mention the content of a postcard that he once sent to a friend enumerating the following New Year's resolutions (cf. [27]):

1. To prove Riemann's Hypothesis,
2. To make a brilliant play in a crucial cricket match,
3. To prove the nonexistence of God,
4. To be the first man atop Mount Everest
5. To be proclaimed the first president of the U.S.S.R., Great Britain and Germany,
6. To murder Mussolini.

In the summer of 1947, Hardy tried to commit suicide by an overdose of barbiturates. However, it was such an exaggerated overdose which was rejected by his body and thus he actually survived. He died on the December 1 1947, in Cambridge, England, at the age of 70.

4.2.3 John Edensor Littlewood

John Edensor Littlewood was born on the June 9 1885, in Rochester, Kent, England. Like Hardy, he was born in a family which was able to provide him mathematical training. However, he was much more privileged since the mathematical heritage of his family was far richer. Not only his father and grandfather were mathematically educated, but they had both participated in the Mathematical Tripos of the University of Cambridge ranking ninth and thirty-fifth, respectively.
In 1892, ten years after his participation at the Mathematical Tripos, Littlewood's father received two important offers: a Fellowship at Magdalene College of the University of Cambridge and the position of Headmaster in a newly founded school at Wynberg, Cape Town, South Africa. He accepted the second offer and the same year the whole family moved to South Africa. Young Littlewood greatly enjoyed the nature and the climate of his new home and thus fondly remembered his childhood there throughout his life. However, his education in Wynberg was poor and his father feared that if his son remained in South Africa, this would severely affect his mathematical training. Hence, in 1900, Littlewood was sent to St. Paul's School, in London, one of the most famous schools for the cultivation of mathematical skills. In his second year of studies there, he won a scholarship to Trinity College, Cambridge. In 1905, while just 19, he participated in the Mathematical Tripos of Cambridge and succeeded to rank first (Senior Wrangler) along with James Mercer. Obtaining this honour was once regarded as the greatest intellectual achievement attainable in Great Britain. At the time, Senior Wranglers were considered to be celebrities in the university and their photographs were sold during May Week, the end of the academic year period which is usually associated with several festivities.
In 1906, Littlewood started conducting research under the supervision of E. W. Barnes. Due to Littlewood's mathematical precocity as well as the success with which he tackled the first problem that was given to him, Barnes decided to propose to him to prove Riemann's Hypothesis.
Apparently, Barnes was unaware of the fact that the Riemann Hypothesis was related to the distribution of prime numbers, a result that was known at his time. Littlewood rediscovered this and managed to give a proof of the prime number theorem under the assumption of the Riemann Hypothesis.

One year after he commenced his research, Littlewood began lecturing at the University of Manchester as Richardson Lecturer, a position he would hold for a bit less than three years after which he returned to Cambridge succeeding A. N. Whitehead. Recollecting the period of his life up to 1910–1911, he once wrote:

"On the looking back this time seems to me to mark my arrival at a reasonably assured judgment and taste, the end of my 'education'. I soon began my 35-year collaboration with Hardy." [1]

Like Hardy, Littlewood's personality had certain peculiarities too. For example, even though he kept very close lifelong friendships, he always wanted these relationships not to affect his complete independence and hence never enjoyed their company for long stretches of time.

For a large part of his life, he suffered from depression and that was one of the reasons why he rarely left Cambridge. Since, evidently, very few people had seen him outside Cambridge to actually acknowledge his existence, there was a joke at the time that Littlewood was a figment of Hardy's imagination.

As an advisor, Littlewood had an interesting way of suggesting research problems to his students. Sir Peter Swinnerton-Dyer, one of Littlewood's students and a famous number theorist, once said:

"He (Littlewood) had a list of twenty or thirty problems, gradually renewed. Some of them looked difficult, and were; others looked easy, and were not. A student could try any he chose, and if he failed on one, he could go on to another ... Each problem was one which a mathematician whom Littlewood respected had seriously attempted and had failed to solve." (cf. [6] for the full quote and for an extensive biography of J. E. Littlewood).

During his life, he was bestowed several awards and was honoured with important positions. In 1916, he was elected a Fellow of the Royal Society, and in 1929, he was awarded the Royal Medal. He was bestowed the De Morgan Medal in 1938, the Sylvester Prize in 1943, the Copley Medal in 1958, and the Senior Berwick Prize in 1960. He also served as a president of the London Mathematical Society from 1941 to 1943. Additionally, he was elected a foreign member of the Göttingen Academy of Sciences and Humanities in 1925, the Swedish Academy and the Royal Danish Academy in 1948, the Dutch Academy in 1950 and the French Academy of Sciences in 1957.

Other than Mathematics, Littlewood loved music and rock climbing. Regarding his music taste, he was devoted to Bach, Beethoven and Mozart. Bella Bollobás, a famous mathematician and a close friend of his, mentions in his

[1]Concerning the Hardy-Littlewood collaboration, the reader is referred to the biography of G. H. Hardy in Sect. 4.2

book *Littlewood's Miscellany* (see [4]) that Littlewood *"considered life too short to waste on other composers"*.
J. E. Littlewood died on the September 6 1977, in Cambridge at the age of 92.

4.2.4 Ivan Matveevich Vinogradov

Ivan Matveevich Vinogradov was born on the September 14 1891, in the village Milolyub in the Velikie Luki, district of the Pskov Province of Russia. Vinogradov showed extraordinary brilliance in a very early age. For example, he learned how to read and write by the age of three. A most surprising thing, though, was that he managed to understand the pattern with which he could pronounce and spell words within a month.
As a son of a priest in Russia of the late nineteenth century, one would expect him to enter an ecclesiastic school and probably one day even become a priest. However, due to his talent in mathematics early in life, in 1903, he was sent to a science school. After graduating in 1910, he entered the Department of Mathematics and Physics of St. Petersburg's University. There he studied under the guidance of Ya. V. Uspenskii, who was his undergraduate thesis advisor. Vinogradov graduated from St. Petersburg's University in 1914. During the same year, he published his first paper on the quadratic residues and nonresidues. Four years after that, he started teaching at the Perm State University (now known as the Gorky State University), where he remained for two years. In 1920, he was elected professor at the Polytechnic Institute of St. Petersburg, and five years later, he was granted a professorship at St. Petersburg's University.
During his life, he was awarded several prestigious prizes and distinctions. Among the most important ones were his election as an honorary member of the London Mathematical Society in 1939, his election as a Fellow of the Royal Society of London in 1942, the invitation to deliver a lecture at the 1966 International Congress of Mathematicians and the receipt of the Lomonosov Gold Medal in 1970, which was the most prestigious award bestowed at that time by the USSR Academy of Sciences.
Besides his great talent in mathematics, it is said that Vinogradov had also a tremendous physical strength and a robust health. Rumour has it that he usually bragged about his strength and, as an example, in 1946 he lifted a concert grand piano at the Royal Society in London!
I. M. Vinogradov died on the 20th of March, 1983, in Moscow at the age of 91.

4.2.5 Lev Genrikhovich Schnirelmann

Lev Genrikhovich Schnirelmann was born on the January 2 1905, in Gomel, Russia. Schnirelmann, like Vinogradov, showed remarkable brilliance at a very early age. For example, he entered the Moscow State University in 1921, when he was just 16 years old. There he studied under the guidance of Nikolai Luzin. He graduated four years later, in 1925. In 1931, he moved to Göttingen, Germany, in order to study at the prestigious University of Göttingen. On the 17th of September of the same year, in one of the meetings of the German Mathematical Society, he presented results in Number Theory, including his remarkable theorem related to Goldbach's Conjecture. Namely, he presented a proof of the fact that there exists a positive integer q such that every integer greater that 1 can be represented as the sum of at most q prime numbers. His results were published in 1933. During the same year, he was elected a member of the Steklov Mathematical Institute.
Schnirelmann died on the March 24 1938, in Moscow at the young age of 34.

4.2.6 Jingrun Chen

Jingrun Chen was born on the May 22 1933, in Fuzhou, the capital of the Fujian Province of China. He was the third son of a large and relatively poor family.
In 1950, after graduating from the high school of his hometown, he entered the Mathematics Department of the University of Xiamen, a city on the southeast coast of China, more than 200 km away from Fuzhou. Three years later, he graduated from the university and was appointed as a teacher at the fourth middle school of Beijing. However, he did not manage to remain in that position for more than a year, since he was fired in 1954. In 1955, he found a job as a clerk at the library of the Mathematics Department of Xiamen University.
Unfortunately, Cold War political tensions were escalated the same year by mainland China, when they bombed some islands held by Taiwan (1955 Taiwan Strait Crisis). This resulted to the bombing of Xiamen, as a response of the Nationalists.
Due to this political climate, air raid alarms were common during Chen's stay in Xiamen. By that time, he had developed such a great passion for number theory that, notwithstanding the difficult times, he carried Loo Keng Hua's "Additive Prime Number Theory" book with him to the shelters, in order to study. The fruit of his research and studies during that period was his paper

entitled “On Tarry’s problem”, which he presented in August 1956 at the Annual meeting of the Chinese Mathematical Society. The next year, after the recommendation of Hua, he joined the Institute of Mathematics of the Academia Sinica.
In 1966, Chen proved the major result related to Goldbach’s Strong Conjecture that for every sufficiently large natural number N, it is true that

$$2N = P_1 + P_2 ,$$

where generally P_m denotes the product of at most m prime numbers.
However, his result was not published until 1973. The reason that his theorem was kept in the dark for so long was that during that period, a sociopolitical movement against capitalism was taking place in China. That movement led to the launching of the so-called Great Proletarian Cultural Revolution in May 1966. During the Cultural Revolution, Chen faced some severe difficulties. Specifically, a high-rank politician blackmailed him in order to sign a document indicting Hua of stealing his mathematical discoveries. This led Chen to an emotional turmoil, which led him to try to commit suicide by jumping out the window of his room. Fortunately, he was not fatally injured.
Soon after the end of the revolution in 1976, a Chinese magazine published Chen’s story which made him nationally well known. In 1978, he was elected as a research professor at the Institute of Mathematics of the Chinese Academy. Chen died on the March 19 1996, at the age of 62.

4.2.7 Harald Andrés Helfgott

Harald Andrés Helfgott was born on the November 25th 1977, in Lima, Perú. Helfgott’s mathematical talent was revealed in an early age, and it is interesting to mention that at the age of 12, he was able to read drafts of his father’s books in Geometry and Introductory Calculus. He then participated in local groups of students in Lima, who were preparing for South American mathematical competitions.
After obtaining his B.A. in Mathematics and Computer Science from Brandeis University in 1998, he entered Princeton University where he did doctoral research under the supervision of Henryk Iwaniec, with Peter Sarnak as his co-advisor. He obtained his Ph.D. from Princeton in 2003. He was a post-doc at Yale and Université de Montréal, and, after a first position at Bristol (UK), he joined the Centre National de la Recherche Scientifique (France) as a full-time researcher. He is currently the Alexander von Humboldt Professor at the University of Göttingen. To date, he has been awarded the Leverhulme

Prize, the Whitehead Prize of the London Mathematical Society and, jointly with Tom Sanders, the Adams Prize of the University of Cambridge. His main areas of work include analytic number theory, diophantine geometry, combinatorics and group theory. He is known, among other reasons, for his work on growth in groups.
In 2013, Helfgott impressed the mathematical community by releasing two papers where he presented an unconditional proof of Goldbach's Ternary Conjecture. Namely, he proved that every odd integer greater than or equal to 7 can be expressed as the sum of three prime numbers [24].

For Further Reading

• G.I. Arhipov, J.Y. Chen, V.N. Chubarikov, On the cardinality of an exceptional set in a binary additive problem of the Goldbach type, in *Proceedings of the Session in Analytic Number Theory and diophantine Equations* (Bonner Math. Schriften 360, Bonn, 2003)
• G.I. Arhipov, V.N. Chubarikov, On the number of summands in Vinogradov's additive problem and its generalizations, in *Modern Problems of Number Theory and its Applications: Current Problems, (Russian)*, vol.1, (Moscow, 2002), pp. 5–38.
• V.M. Arkhangelskaya, Some calculations connected with Goldbach's problem. Ukraine Math. J. **9**, 20–29 (1957)
• R.C. Baker, G. Harman, The three primes theorem with almost equal summands. R. Soc. Lond. Philos. Trans. Ser. A **356**, 763–780 (1998)
• K.G. Borodzkin, On I. M. Vinogradov's constant, in *Proceedings of 3rd All-Union Math. Conference*, vol. 1, Izdat. Akad. Nauk SSSR, (Moscow, 1956).
• J. Brüdern, K. Kawada, Ternary problems in additive prime number theory, in *Analytic Number Theory* (Kluwer Academic Publishers, Dordrecht, 2002), pp. 39–91
• J.W.S. Cassels, R.C. Vaughan, Ivan Matveevich Vinogradov. Biographical Memoirs of Fellows of the Royal Society, **31**, 613–629 (1985)
• J.R. Chen, On large odd numbers as sum of three almost equal primes. Sci. Sinica **14**, 1113–1117 (1965)
• J.R. Chen, C.D. Pan, The exceptional set of Goldbach numbers. Sci. Sinica **23**, 416–430 (1980)
• N.G. Chudakov, On the Goldbach problem. C. R. Acad. Sci. URSS. **17**(2), 335–338 (1937)
• N.G. Chudakov, On the density of the sets of even integers which are not representable as a sum of two odd primes, (Russian), Izv. Akad. Nauk SSSR Ser. Mat. **2**, 25–40 (1938)

M.T. Rassias, *Goldbach's Problem*, DOI 10.1007/978-3-319-57914-6

• N.G. Chudakov, On the Goldbach–Vinogradov's theorem, Ann. Math. **48**(2), 515–545 (1947)
• J.M. Deshouillers, G. Effinger, H. te Riele, D. Zinoviev, A complete Vinogradov 3-primes theorem under the Riemann hypothesis, Electron. Research Announcements Am. Math. Soc. **3**, 99–104 (1997)
• T. Estermann, On Goldbach's problem: Proof that almost all even positive integers are sums of two primes. Proc. Lond. Math. Soc. Ser. **44**(2), 307–314 (1938)
• T. Estermann, Proof that every large integer is the sum of two primes and a square. Proc. Lond. Math. Soc. **34**(2), 190–195 (1932)
• B. Green, T. Tao, The primes contain arbitrarily long arithmetic progressions. Ann. Math. **167**(2), 481–547 (2008)
• G.H. Hardy, Goldbach's theorem. Math. Tid. B, 1–16 (1922)
• D.R. Heath-Brown, A new form of the circle method, and its application to quadratic forms. J. Reine Angew. Math. **481**, 149–206 (1996)
• D.R. Heath-Brown, The circle method and diagonal cubic forms. R. Soc. Lond. Philos. Trans. Ser. A. **356**, 673–699 (1998)
• L.K. Hua, Some results in additive prime number theory. Quart. J. Math. Oxford **9**, 68–80 (1938)
• L.K. Hua, *Additive Theory of Prime Numbers*. (Americal Mathematical Society, Providence, RI, 1965)
• C.H. Jia, The three primes theorem over short intervals, (Chinese), Acta Math. Sinica, **32**, 464–473 (1989)
• A.A. Karatsuba, I. M. Vinogradov and his method of trigonometric sums, in *Proceedings of the Steklov Institute of Mathematics*. Number Theory and Analysis, vol. 207, no. 6 (1995), pp. 1–25
• A. Languasco, On the exceptional set of Goldbach's problem in short intervals. Monatsh. Math. **141**, 147–169 (2004)
• A.F. Lavrik, On the representation of numbers as the sum of primes by Schnirel'man's method. Izv. Akad. Nauk UzSSR Ser. Fiz. Mat. Nauk **3**, 5–10 (1962)
• J. Law, *The Circle Method on the Binary Goldbach Conjecture*, (Princeton University, 2005)
• H.Z. Li, Prime numbers with powers of two, (Russian), Trudy Mat. Inst. Steklov **38**, 152–169 (1951)
• Yu.V. Linnik, A new proof of the Goldbach–Vinogradov theorem. Mat. Sb., **19**(61), 3–8 (1946)
• Yu.V. Linnik, Prime numbers and powers of two. Trudy Mat. Inst. Steklov **38**, 151–169 (1951)
• Yu.V. Linnik, Addition of prime numbers and powers of one and the same number, Mat. Sb. (N. S.) **32**, 3–60 (1953)

• B. Lucke, *Zur Hardy-Littlewoodschen Behandlung des Goldbachschen Problems*, (Doctoral Dissertation, University of Göttingen, 1926)
• A.P. Lursmanashvili, *Representation of natural numbers by sums of prime numbers,* Thbilis. Sahelmc. Univ. Shrom. Mekh.-Math. Mecn. Ser. **117**, 63–76 (1966)
• H.L. Montgomery, Problems concerning prime numbers. Proc. Symposia Pure Math. **28**, 307–310 (1976)
• M.B. Nathanson, *Additive Number Theory: The Classical Bases*, (Springer-Verlag, 1996)
• C.B. Pan, C.D. Pan, *Goldbach Conjecture*, (Science Press, 1992)
• T.P. Peneva, On the ternary Goldbach problem with primes p_i *such that* $p_i + 2$ are almost prime, Acta Math. Hungar. **86**, 305–318 (2000)
• X. Ren, The Waring–Goldbach problem for cubes, Acta Arith. **94**, 287–301 (2000)
• A. Rényi, On the representation of an even number as a sum of a single prime and a single almost prime number, (Russian), Dokl. Akad. Nauk SSSR, **56**, 455–458 (1947)
• A. Selberg, On an elementary method in the theory of primes, Norske Vid. Selsk. Forh. Trondhjem, **19**, 64–67 (1947)
• R.C. Vaughan, A new estimate for the exceptional set in Goldbach's problem. Am. Math. Soc. Proc. Symp. Pure Math. **24**, 315–320 (1973)
• R.C. Vaughan, A survey of recent work in additive prime number theory, Sem. Théor. Nombres, Bordeaux **19**, 1–7 (1973/1974)
• R.C. Vaughan, Hardy's Legacy to number theory. J. Austral. Math. Soc. Ser. A **65**, 238–266 (1998)
• R.C. Vaughan, A ternary additive problem. Proc. Lond. Math. Soc. **41**, 516–532 (1980)
• R.C. Vaughan, Recent work in additive prime number theory, in *Proceedings of the International Congress of Mathematicians* (Helsinki, 1978), pp. 389–394
• I.M. Vinogradov, *The Method of Trigonometrical Sums in the Theory of Numbers*, (Interscience Publ., London, New York, 1954)
• A.I. Vinogradov, The binary Hardy–Littlewood problem (Russian). Acta Arith. **46**, 33–56 (1985)
• T.Z. Wang, On Linnik's almost Goldbach theorem. Sci. China Ser. A **42**, 1155–1172 (1999)
• A.P. Yushkevich, L. G. Shnirel'man in Göttingen, (Russian). Istor. Mat. Issled. **28**, 287–290, 351 (1985)
• T. Zhan, On the representation of large odd integer as a sum of three almost equal primes, Acta Math. Sinica (N.S.) **7**, 259–272 (1991)

Bibliography

1. M. Balazard, Elementary remarks on Möbius' function, in *Proceedings of the Steklov Intitute of Mathematics*, (2012), pp. 276
2. A. Balog, J. Friedlander, A hybrid of theorems of Vinogradov and Piatetski-Shapiro. Pac. J. Math. **156**(1), 45–62 (1992)
3. A. Balog, A. Perelli, Exponential sums over primes in an arithmetic progression. Proc. Am. Math. Soc. **93**, 578–582 (1985)
4. B. Bollobás (ed.), *Littlewood's Miscellany* (Cambridge University Press, Cambridge, New York, 1986)
5. N.G. de Bruijn, On the number of positive integers $\leq x$ and free of prime factors $>y$. Nederl. Acad. Wetensch. Proc. Ser. A **54**, 50–60 (1951)
6. J.C. Burkill, John Edensor Littlewood. Bull. Lond. Math. Soc. **11**, 59–103 (1979)
7. J.R. Chen, On the representation of large even integer as the sum of a prime and the product of at most two primes. Sci. Sinica **16**, 157–176 (1973)
8. H. Daboussi, J. Rivat, Explicit upper bounds for exponential sums over primes. Math. Comp. **70**(233), 431–447 (2001)
9. H. Davenport, *Multiplicative Number Theory* (Springer, New York, 1980)
10. J.M. Deshouillers, *Sur la constante de Šnirel'man*, In: Séminaire Delange-Pisot-Poitou, 17e année: (1975/1976), Théorie des nombres: Fac. 2, Exp. No. G16, p. 6. Secrétariat Math., Paris (1977)
11. F. Dress, H. Iwaniec, G. Tenenbaum, Sur une somme liée à la fonction de Möbius. J. Reine Angew. Math. **340**, 53–58 (1983)
12. P. Erdős, On the difference of consecutive primes. Q. J. Math. Oxf. Ser. **6**, 124–128 (1935)
13. P. Erdős, M.B. Nathanson, Lagrange's Theorem and thin subsequences of squares, in *Contributions to Probability*, eds. by J. Gani, V.K. Rohatgi (Academic Press, New York, 1981), pp. 3–9
14. T. Estermann, *Introduction to Modern Prime Number Theory* (Cambridge University Press, Cambridge, 1961)
15. K. Ford, B. Green, S. Konyagin, T. Tao, *Large gaps between consecutive prime numbers*. preprint, arXiv: 1408.4505
16. K. Ford, B. Green, S. Konyagin, J. Maynard, T. Tao, *Long gaps between primes*, arXiv:1412.5029v3
17. B. Green, Generalizing the Hardy-Littlewood method for primes, in *Proceedings of the International Congress of Mathematicians, Madrid*, (2006), pp. 373–399. (European Math. Soc., Vol. II)

M.T. Rassias, *Goldbach's Problem*, DOI 10.1007/978-3-319-57914-6

18. H. Halberstam, H.-E. Richert, *Sieve Methods* (Academic Press, London, 1974)
19. G.H. Hardy, J.E. Littlewood, Some problems of "Partitio Numerorum". III. On the expression of a number as a sum of primes. Acta Math. **44**, 1–70 (1923)
20. G.H. Hardy, J.E. Littlewood, Some problems of "Partitio Numerorum". V. A further contribution to the study of Goldbach's problem, Proc. London Math. Soc. **2**(22), 46–56 (1923)
21. G.H. Hardy, S. Ramanujan, Asymptotic formulae in combinatorial analysis. Proc. Lond. Math. Soc. **17**, 75–115 (1918)
22. D.R. Heath-Brown, Three primes and an almost–prime in arithmetic progression, J. Lond. Math. Soc. **2**(23), 396–414 (1981)
23. D.R. Heath-Brown, The ternary Goldbach problem. Rev. Mat. Iberoamericana **1**(1), 45–59 (1985)
24. H.A. Helfgott, *Major arcs for Goldbach's theorem*, Annals of Mathematics Studies, Princeton, to appear
25. H.A. Helfgott, D.J. Platt, Numerical verification of the ternary Goldbach conjecture up to $8.875 \cdot 10^{30}$. Exp. Math. **22**(4), 406–409 (2013)
26. D. Hilbert, Mathematical problems. Lecture delivered before the International Congress of Mathematicians at Paris in 1900, Bull. Am. Math. Soc. **8**, 437–479 (1902)
27. P.H. Hoffman, *The Man Who Loved Only Numbers: The Story of Paul Erdős and the Search for Mathematical Truth* (Hyperion, New York, 1998)
28. L.K. Hua, *Introduction to Number Theory*, (Springer, Berlin, 1982)
29. H. Iwaniec, E. Kowalski, *Analytic Number Theory*, vol. 53, (American Mathematical Society Colloquium Publications, Providence, RI, 2004)
30. H. Iwaniec, *Lectures on the Riemann Zeta Function*, University Lecture Series, vol. 62 (American Mathematical Society, Providence, RI, 2014)
31. H. Kadiri, A. Lumley, Short effective intervals containing primes, Submitted (2014), arXiv:1407:7902
32. R. Kanigel, *The Man Who Knew Infinity: A Life of the Genius Ramanujan* (Charles Scribner's Sons, New York, 1991)
33. A.V. Kumchev, D.I. Tolev, An invitation to additive prime number theory, Serdica Math. J. **31**(1–2), 1–74 (2005), http://front.math.ucdavis.edu/0412.5220
34. J. Liu, T. Zhan, The ternary Goldbach problem in arithmetic progressions. Acta Arith. **82**(3), 197–227 (1997)
35. M.C. Liu, T.Z. Wang, On the Vinogradov bound in the three primes Goldbach conjecture. Acta Arith. **105**, 133–175 (2002)
36. H. Maier, C. Pomerance, Unusually large gaps between consecutive primes. Trans. Am. Math. Soc. **322**(1), 201–237 (1990)
37. H. Maier, M.Th. Rassias, The ternary Goldbach problem with a prime and two isolated primes, in *Proceedings of the Steklov Institute of Mathematics (to appear)*
38. J. Maynard, *Large gaps between primes*, preprint, arXiv:1408.5110
39. K.S. McCurley, Explicit estimates for the error term in the prime number theorem for arithmetic progressions. Math. Comp. **42**, 265–285 (1984)
40. S.J. Miller, R. Takloo-Bighash, *An Invitation to Modern Number Theory* (Princeton University Press, Princeton and Oxford, 2006)
41. H.L. Montgomery, R.C. Vaughan, The exceptional set in Goldbach's problem. Acta Arith. **27**, 353–370 (1975)
42. L.I. Piatetski-Shapiro, On the distribution of prime numbers in sequences of the form $[f(m)]$. Math. Sb. **33**, 559–566 (1953)
43. W.W. Pin, *Vinogradov's Theorem and its Generalization on Primes in Arithmetic Progression*, (National University of Singapore, 2009)
44. J. Pintz, Very large gaps between consecutive primes. J. Number Theory **63**(2), 286–301 (1997)
45. D.J. Platt, Numerical computations concerning the GRH, Ph.D. thesis, 2013, arXiv:1305.3087
46. H. Rademacher, Uber eine Erweiterung des Goldbachschen Problems. Math. Zeitschrift **25**, 627–657 (1926)
47. O. Ramaré, On Šnirel'man's constant, Ann. Scuola Norm. Sup. Pisa **4**(22), 645–706 (1995)

48. O. Ramaré, *Arithmetical aspects of the large sieve inequality*, vol. 1 of Harish-Chandra Research Institute Lecture Notes, (Hindustan Book Agency, New Delhi, 2009) With the collaboration of D. S. Ramana
49. O. Ramaré, On Bombieri's asymptotic sieve. J. Number Theory **130**(5), 1155–1189 (2010)
50. O. Ramaré, From explicit estimates for the primes to explicit estimates for the Moebius function. Acta Arith. **157**(4), 365–379 (2013)
51. O. Ramaré, Explicit estimates on several summatory functions involving the Moebius function. Math. Comp. **84**, 1359–1387 (2015)
52. O. Ramaré, I.M. Ruzsa, Additive properties of dense subsets of sifted sequences. J. Théorie N. Bordeaux **13**, 559–581 (2001)
53. R.A. Rankin, The difference between consecutive prime numbers. J. Lond. Math. Soc. **13**, 242–247 (1938)
54. R. A. Rankin, *The difference between consecutive prime numbers. V.* Proc. Edinburgh Math. Soc., 12/13 (1962/63), 331–332
55. MTh Rassias, *Problem-Solving and Selected Topics in Number Theory: In the Spirit of the Mathematical Olympiads* (Springer, New York, 2011)
56. H. Riesel, R.C. Vaughan, On sums of primes. Ark. Mat. **21**(1), 46–74 (1983)
57. J.B. Rosser, Explicit bounds for some functions of prime numbers. Am. J. Math. **63**, 211–232 (1941)
58. L.G. Schnirelmann, Über additive Eigenschaften von Zahlen. Math. Ann. **107**, 649–690 (1932/1933)
59. A. Schönhage, Eine Bemerkung zur Konstruktion grosser Primzahllücken. Arch. Math. **14**, 29–30 (1963)
60. T. Tao, Every odd number greater than 1 is the sum of at most five primes. Math. Comp. **83**(286), 997–1038 (2014)
61. R.C. Vaughan, On Goldbach's problem. Acta Arith. **22**, 21–48 (1972)
62. R.C. Vaughan, On the estimation of Schnirelman's constant. J. Reine Angew. Math. **290**, 93–108 (1977)
63. R.C. Vaughan, Sommes trigonométriques sur les nombres premiers, C. R. Acad. Sci. Paris Sér. A-B **285**(16), A981–A983 (1977)
64. R.C. Vaughan, *The Hardy-Littlewood Method*, (Cambridge University Press, Cambridge, 1981)
65. R.C. Vaughan, Goldbach's conjectures: A historical perspective, in *Open Problems in Mathematics*, eds. J.F. Nash, Jr., M.Th. Rassias, (Springer, New York, 2016), pp. 479–520
66. I.M. Vinogradov, Representation of an odd number as the sum of three primes. Dokl. Akad. Nauk SSSR **15**, 291–294 (1937)
67. I.M. Vinogradov, Some theorems concerning the theory of primes. Mat. Sbornik N. S. **2**, 179–195 (1937)
68. E. Westzynthius, Über die Verteilung der Zahlen die zu den *n* ersten Primzahlen teilerfremd Sind. Commun. Phys. Math. Helsingfors **25**, 1–37 (1931)
69. E. Wirsing, Thin subbases. Analysis **6**, 285–308 (1986)

Index

M.T. Rassias, *Goldbach's Problem*, DOI 10.1007/978-3-319-57914-6

Printed in the United States
By Bookmasters

Printed in the United States
By Bookmasters